# Lecture Notes in Computer Science

Vol. 219: Advances in Cryptology – EUROCRYPT '85. Proceedings, 1985. Edited by F. Pichler. IX, 281 pages. 1986.

Vol. 220: RIMS Symposia on Software Science and Engineering II. Proceedings, 1983 and 1984. Edited by E. Goto, K. Araki and T. Yuasa. XI, 323 pages. 1986.

Vol. 221: Logic Programming '85. Proceedings, 1985. Edited by E. Wada. IX, 311 pages. 1986.

Vol. 222: Advances in Petri Nets 1985. Edited by G. Rozenberg. VI, 498 pages. 1986.

Vol. 223: Structure in Complexity Theory. Proceedings, 1986. Edited by A.L. Selman. VI, 401 pages. 1986.

Vol. 224: Current Trends in Concurrency. Overviews and Tutorials. Edited by J.W. de Bakker, W.-P. de Roever and G. Rozenberg. XII, 716 pages. 1986.

Vol. 225: Third International Conference on Logic Programming. Proceedings, 1986. Edited by E. Shapiro. IX, 720 pages. 1986.

Vol. 226: Automata, Languages and Programming. Proceedings, 1986. Edited by L. Kott. IX, 474 pages. 1986.

Vol. 227: VLSI Algorithms and Architectures – AWOC 86. Proceedings, 1986. Edited by F. Makedon, K. Mehlhorn, T. Papatheodorou and P. Spirakis. VIII, 328 pages. 1986.

Vol. 228: Applied Algebra, Algorithmics and Error-Correcting Codes. AAECC-2. Proceedings, 1984. Edited by A. Poli. VI, 265 pages. 1986.

Vol. 229: Algebraic Algorithms and Error-Correcting Codes. AAECC-3. Proceedings, 1985. Edited by J. Calmet. VII, 416 pages. 1986.

Vol. 230: 8th International Conference on Automated Deduction. Proceedings, 1986. Edited by J.H. Siekmann. X, 708 pages. 1986.

Vol. 231: R. Hausser, NEWCAT: Parsing Natural Language Using Left-Associative Grammar. II, 540 pages. 1986.

Vol. 232: Fundamentals of Artificial Intelligence. Edited by W. Bibel and Ph. Jorrand. VII, 313 pages. 1986.

Vol. 233: Mathematical Foundations of Computer Science 1986. Proceedings, 1986. Edited by J. Gruska, B. Rovan and J. Wiedermann. IX, 650 pages. 1986.

Vol. 234: Concepts in User Interfaces: A Reference Model for Command and Response Languages. By Members of IFIP Working Group 2.7. Edited by D. Beech. X, 116 pages. 1986.

Vol. 235: Accurate Scientific Computations. Proceedings, 1985. Edited by W.L. Miranker and R.A. Toupin. XIII, 205 pages. 1986.

Vol. 236: TEX for Scientific Documentation. Proceedings, 1986. Edited by J. Désarménien. VI, 204 pages. 1986.

Vol. 237: CONPAR 86. Proceedings, 1986. Edited by W. Händler, D. Haupt, R. Jeltsch, W. Juling and O. Lange. X, 418 pages. 1986.

Vol. 238: L. Naish, Negation and Control in Prolog. IX, 119 pages. 1986.

Vol. 239: Mathematical Foundations of Programming Semantics. Proceedings, 1985. Edited by A. Melton. VI, 395 pages. 1986.

Vol. 240: Category Theory and Computer Programming. Proceedings, 1985. Edited by D. Pitt, S. Abramsky, A. Poigné and D. Rydeheard. VII, 519 pages. 1986.

Vol. 241: Foundations of Software Technology and Theoretical Computer Science. Proceedings, 1986. Edited by K.V. Nori. XII, 519 pages. 1986.

Vol. 242: Combinators and Functional Programming Languages. Proceedings, 1985. Edited by G. Cousineau, P.-L. Curien and B. Robinet. V, 208 pages. 1986.

Vol. 243: ICDT '86. Proceedings, 1986. Edited by G. Ausiello and P. Atzeni. VI, 444 pages. 1986.

Vol. 244: Advanced Programming Environments. Proceedings, 1986. Edited by R. Conradi, T.M. Didriksen and D.H. Wanvik. VII, 604 pages. 1986

Vol. 245: H.F. de Groote, Lectures on the Complexity of Bilinear Problems. V, 135 pages. 1987.

Vol. 246: Graph-Theoretic Concepts in Computer Science. Proceedings, 1986. Edited by G. Tinhofer and G. Schmidt. VII, 307 pages. 1987.

Vol. 247: STACS 87. Proceedings, 1987. Edited by F.J. Brandenburg, G. Vidal-Naquet and M. Wirsing. X, 484 pages. 1987.

Vol. 248: Networking in Open Systems. Proceedings, 1986. Edited by G. Müller and R.P. Blanc. VI, 441 pages. 1987.

Vol. 249: TAPSOFT '87. Volume 1. Proceedings, 1987. Edited by H. Ehrig, R. Kowalski, G. Levi and U. Montanari. XIV, 289 pages. 1987.

Vol. 250: TAPSOFT '87. Volume 2. Proceedings, 1987. Edited by H. Ehrig, R. Kowalski, G. Levi and U. Montanari. XIV, 336 pages. 1987.

Vol. 251: V. Akman, Unobstructed Shortest Paths in Polyhedral Environments. VII, 103 pages. 1987.

Vol. 252: VDM '87. VDM – A Formal Method at Work. Proceedings, 1987. Edited by D. Bjørner, C.B. Jones, M. Mac an Airchinnigh and E.J. Neuhold. IX, 422 pages. 1987.

Vol. 253: J.D. Becker, I. Eisele (Eds.), WOPPLOT 86. Parallel Processing: Logic, Organization, and Technology. Proceedings, 1986. V, 226 pages. 1987.

Vol. 254: Petri Nets: Central Models and Their Properties. Advances in Petri Nets 1986, Part I. Proceedings, 1986. Edited by W. Brauer, W. Reisig and G. Rozenberg. X, 480 pages. 1987.

Vol. 255: Petri Nets: Applications and Relationships to Other Models of Concurrency. Advances in Petri Nets 1986, Part II. Proceedings, 1986. Edited by W. Brauer, W. Reisig and G. Rozenberg. X, 516 pages. 1987.

Vol. 256: Rewriting Techniques and Applications. Proceedings, 1987. Edited by P. Lescanne. VI, 285 pages. 1987.

Vol. 257: Database Machine Performance: Modeling Methodologies and Evaluation Strategies. Edited by F. Cesarini and S. Salza. X, 250 pages. 1987.

Vol. 258: PARLE, Parallel Architectures and Languages Europe. Volume I. Proceedings, 1987. Edited by J.W. de Bakker, A.J. Nijman and P.C. Treleaven. XII, 480 pages. 1987.

Vol. 259: PARLE, Parallel Architectures and Languages Europe. Volume II. Proceedings, 1987. Edited by J.W. de Bakker, A.J. Nijman and P.C. Treleaven. XII, 464 pages. 1987.

Vol. 260: D.C. Luckham, F.W. von Henke, B. Krieg-Brückner, O. Owe, ANNA, A Language for Annotating Ada Programs. V, 143 pages. 1987.

Vol. 261: J. Ch. Freytag, Translating Relational Queries into Iterative Programs. XI, 131 pages. 1987.

Vol. 262: A. Burns, A.M. Lister, A.J. Wellings, A Review of Ada Tasking. VIII, 141 pages. 1987.

Vol. 263: A.M. Odlyzko (Ed.), Advances in Cryptology – CRYPTO '86. Proceedings. XI, 489 pages. 1987.

Vol. 264: E. Wada (Ed.), Logic Programming '86. Proceedings, 1986. VI, 179 pages. 1987.

Vol. 265: K.P. Jantke (Ed.), Analogical and Inductive Inference. Proceedings, 1986. VI, 227 pages. 1987.

Vol. 266: G. Rozenberg (Ed.), Advances in Petri Nets 1987. VI, 451 pages. 1987.

Vol. 267: Th. Ottmann (Ed.), Automata, Languages and Programming. Proceedings, 1987. X, 565 pages. 1987.

Vol. 268: P.M. Pardalos, J.B. Rosen, Constrained Global Optimization: Algorithms and Applications. VII, 143 pages. 1987.

Vol. 269: A. Albrecht, H. Jung, K. Mehlhorn (Eds.), Parallel Algorithms and Architectures. Proceedings, 1987. Approx. 205 pages. 1987.

# Lecture Notes in Computer Science

Edited by G. Goos and J. Hartmanis

311

G. Cohen P. Godlewski (Eds.)

# Coding Theory and Applications

2nd International Colloquium
Cachan-Paris, France, November 24–26, 1986
Proceedings

Springer-Verlag
Berlin Heidelberg New York London Paris Tokyo

**Editors**

Gérard Cohen
Département Informatique
Ecole Nationale Supérieure des Télécommunications
46, rue Barrault, F-75634 Paris Cedex 13, France

Philippe Godlewski
Département Réseaux
Ecole Nationale Supérieure des Télécommunications
46, rue Barrault, F-75634 Paris Cedex 13, France

CR Subject Classification (1987): E.4, G.2.1, F.2.1

ISBN 3-540-19368-5 Springer-Verlag Berlin Heidelberg New York
ISBN 0-387-19368-5 Springer-Verlag New York Berlin Heidelberg



Printing and binding: Druckhaus Beltz, Hemsbach/Bergstr.
2145/3140-543210

# Preface

The colloquium *"Trois Journées sur le Codage"* , held in Cachan near Paris, France from $24^{th}$ to $26^{th}$ November 1986, was the second one of this type. It aimed at gathering together approximately one hundred scientists and engineers, mostly from France. With its broad spectrum, ranging from algebraic geometry to implementation of coding algorithms, it was a unique opportunity for contact between university and industry on the topics of information and coding theory.

It is a great pleasure to acknowledge the efforts of Professor Goutelard who played a great part in making the organization of these "3 journées" a success.

We would like to thank the organizing institutions for their help: DRET, LETTI and BULL s.a., the Advisory Board, the Scientific Committee, the Sponsoring Committee, Editors and referees of Springer-Verlag, and our own referees (see "List of Referees") with a special mention for Gilles Zemor.

These contributed papers allow a survey of "hexagonal" research (cf. fig. 1) in coding. The purpose of this introduction is to provide a quick first visit. It has its drawbacks as do most tourist guides: simplifications or information of a local or anecdotal type ... but it mainly aims at giving a taste of French production, in a area different from cookery or wine.

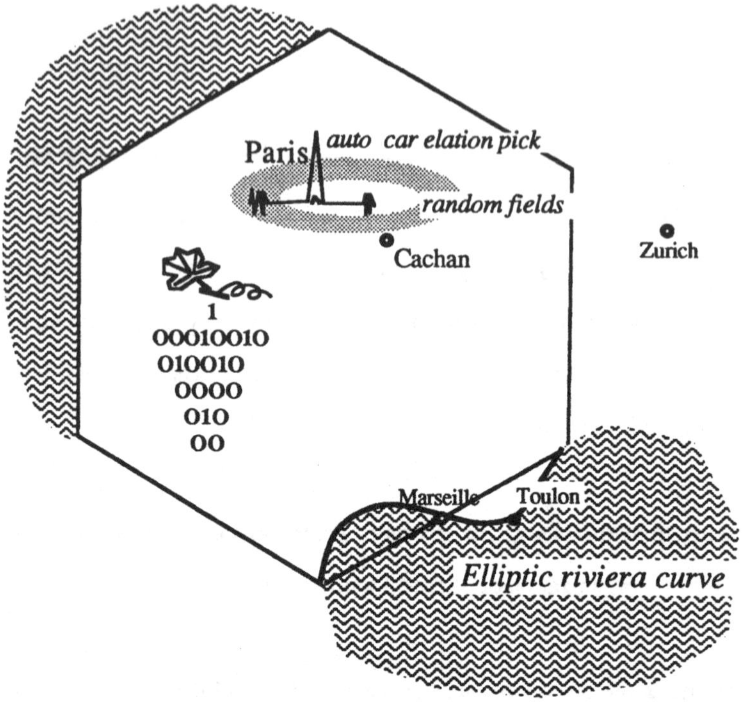


Fig. 1 : French Hexagon

The proceedings begin with the texts of two "almost French" invited speakers: Toby Berger, at that time on sabbatical at ENST, and James Massey, from ETH, Zürich.

The first invited address was on "Information Theory in Random Fields". A random field is a family of random variables with multidimensional parameter set. Random fields provide mathematical models for distributed sources of information. Channels that link an input and an output random field are also of interest.

The second one, entitled "Linear Complexity in Coding Theory", connected different problems: Blahut's Theorem on algebraic decoding, linear complexity of periodic sequences and minimum distance bounds of cyclic codes. These enlightening connections allow the various generalizations of the BCH bound to be gathered into a unified presentation .

## 1. CODING AND COMBINATORICS

Links between Coding and Combinatorics are well known now, e.g. between designs and perfect codes. The goal of this session was to further investigate them.

The paper by G. Roux is an extension of a "design-like" property from sequences over a finite alphabet to permutations. The problem considered here is of covering type: find the minimum number p(n,k) of rows in an array with n columns and elements from {1,2,...n} such that every extracted subarray of k columns contains as rows the k! permutations of the column indices. In particular, for k fixed and n large enough, the order of magnitude of p(n,k) is found to be $n^{k-1}$.

A more classical covering problem is studied in the paper by M. Beveraggi and G. Cohen, namely the minimal possible density $\mu_n$ for a covering of the binary n-dimensional Hamming space by spheres of radius one. It is shown that $\mu_n \leq 3/2$. Another fruitful interaction between Coding and Combinatorics occurred through algebra and association schemes. Initiated by P. Delsarte (see reference 9 in P. Solé's paper), it is extended from the Hamming case to the Lee one by P. Solé, with a generalization of Lloyd's theorem.

Another step off the classical Hamming path in Coding is taken by A. Lobstein in his paper "On Modular Weights in Arithmetic Codes". These codes are used for detecting and correcting errors in computer computations. The weight of these errors can be measured by two types of "modular distances" which do not necessarily satisfy the triangular inequality. The author investigates when these distances coincide.

## 2. VARIABLE LENGTH CODES

Two schools of unequal importance coexist in France. They differ from the usual approach, based on source coding problems, which can be found in most books on information theory.

The first one studies "questionnaires", following C.F. Picard. The point of view is often pragmatic and does not exclude probabilities.

The second one deals with combinatorics on the free monoid, as considered by Schützenberger; a recent book "Theory of Codes" (cf. reference 2 in A. Bertrand's paper) is characteristic of that trend. The approach is mathematical. In fact "Theory of Codes" deals with variable length codes endowed with algebraic, especially non-commutative, structures. The book has very little in common with error-correcting codes, even with "algebraic theory" as considered by, e.g. Berlekamp.

The first article by Toby Berger and Raymond Yeung studies "Optimum 1-Ended Binary Prefix Codes". Such codes can find applications for conflict resolution in a multiple access protocol where active stations are identified by a sequence of answers to queries, represented by a binary variable: 0 no transmission, 1 at least one transmission. To solve the conflict in an efficient way, the last answer should be a "1" corresponding to a successful transmission. A last "0" answer would be a revealing but time-wasting silence.

The second article by H. Akdag and B. Bouchon deals with right leaning trees (or prefix coding) that allow terminal nodes (code words) to be ordered according to their probabilities.

The paper by A. Bertrand investigates the noiseless source coding problem (i.e. how to minimize the average length for a source with given entropy) in a non-classical setting, which is connected to those emerging from constraint channel or line code problems.

## 3. CODING AND ALGEBRAIC GEOMETRY

Quite recently, concepts of algebraic geometry have been successfully applied to Coding, starting with the following very general idea due to Goppa (reference G in Le Brigand): take a projective algebraic curve X over a finite field K, i.e. an algebraic variety of dimension one in the projective m-space $P^m(K)$ over K; choose n rational points (i.e. with coordinates in K) on X, say $P_1, P_2, \dots P_n$, and another rational point Q. For a fixed integer t, consider the linear space L of rational functions f(.) such that the pole of f(.) in Q is at most t. Then

$$C = \{ ( f(P_1), f(P_2), \dots, f(P_m) ), f \in L\}$$

is a linear code over K. Its dimension and minimum distance can be estimated with the Riemann-Roch theorem. This very powerful technique has produced a large class of codes. Some of them lie above the Varshamov-Gilbert bound (see references 11 and 12 in the paper by Michon and Driencourt). For example, the classical definition of Reed-Solomon Codes as

$$E = \{ (f(\alpha_0), f(\alpha_1) \dots f(\alpha_{q-1})), f \in M\},$$

where $\{ \alpha_0, \alpha_1, \dots, \alpha_{q-1}\} = F_q$ and M = {polynomials over $F_q$ of degree at most k-1}, can be put into the framework of Goppa codes as follows: take for X the projective line over $F_q$, with $P_i = ( \alpha_i, 1)$, $0 \leq i \leq q-1$, and point Q being the point at infinity (1,0). Then choose for L the set of rational functions with no poles on the $P_i$'s and a pole of order at most k-1 on Q.

A few extensions of these codes are studied in this book:

- One idea is to replace the curve X in Goppa codes by a variety of higher dimension (e.g. projective spaces): this in done in Lachaud's paper, where he gets projective $r^{th}$ order Reed-Muller codes, attaining Plotkin bound for r = 1.
- Starting with the singular curve $F(X,Y,Z) = Y^2 Z^3+YZ^4+X^5 = 0$ over $F_{16}$, Le Brigand gets a code with parameters [n=32, k=17, d=14].

- One of Goppa's constructions involved the Hermitian curves $X_0^{s+1} + X_1^{s+1} + X_2^{s+1}$ over $PG(2,s^2)$. By embedding it for s = 2 in PG(5,4), this gave a [9,3,6] code over $F_4$.

- By embedding the Hermitian surface $X_0^3 + X_1^3 + X_2^3 + X_3^3 = 0$ of PG(3,4) in PG (9,4), Chakravarti obtains a [45,35,4] code over $F_4$.

- One goal was to improve the tables of best known codes by ad hoc constructions like concatenation. This is done in detail by Driencourt and Michon, following Barg, Katsman and Tsfasman: they use outer geometric codes on elliptic curves over $F_8$ or $F_{16}$ concatenated with inner binary codes. For example, one can get a [104,32,35] binary code, decreasing the best known redundancy for such n and d by 2.

## 4. DECODING IN REAL SPACE

In many engineering applications, it is important to embed discrete space into real space. This can be done to improve performances of a decoding process (cf. soft or weighted decoding algorithms) or in the conception of communication systems with high spectral efficiency.

Since Ungerboeck's work, it is customary to consider codes matched to specific constellations of points in real space. These codes are adapted to Viterbi decoding which operates on a trellis. In an article "A Lower Bound on the Minimum Euclidian Distance of Trellis Codes", Marc Rouanne and Daniel J. Costello obtain a random coding bound over the set of non-linear time-varying trellis codes, which provides a means of comparing asymptotic performances of different modulation schemes.

Soft decoding algorithms have been studied for more than 25 years. With respect to the general complexity of this problem, no significant breakthrough, comparable to the

Berlekamp-Massey algorithm in the hard decoding case, has been made for a large class of codes. For general linear codes the problem is NP-complete as underlined by J. Fang et al. An interesting approach initialized by G. Battail consists of using cross-entropy minimization under the following artificial constraint: the a posteriori probability distribution is separable (i.e. the random variables corresponding to the decoded symbols are independent). Such a method can be of interest in a concatenated scheme where it is important to have weighted symbols at the input of the second decoder. The complete paper will appear in French in *Annales des Télécommunications*. J.C. Belfiore, using the same assumption on the a posteriori distribution, obtains a decoding algorithm for binary convolutional codes which belongs to the same family.

## 5. APPLICATIONS

Roger Alexis studies sequences with periodic autocorrelation equal to zero on 2n points round the central peak. Such sequences, called CAZAC (for Constant Amplitude and Zero Autocorrelation), are used for synchronization purposes or for estimating the transfer function of a transmission channel by sounding. The author obtains by a shortened exhaustive search all binary sequences for small length N, $N \leq 20$ and for $N = 32$, and quaternary sequences for $N \leq 12$. To transmit information over a deeply distorted and noisy channel like an underwater channel, one can use a set of sequences with suitable correlation properties. For such a channel, Gaspard Hakizimana, G. Jourdain and G. Loubet compare two signalling sets. The first one, called pseudo-orthogonal coset code, consists of sequences with small cross-correlation and out-of-phase autocorrelation (e.g. Gold Sequences). The second one, due to Yates and Holgate, involves sequences with multi-peak autocorrelation; it is more sensitive to the distribution of channel path delays but leads to simpler receivers.

Francisco Garcia-Ulgade compares 3 variants of the algebraic decoding method and gives the processing time of the corresponding algorithms implemented on a 16-bit microprocessor (M68000) for [15,9,7] and [31,25,7] Reed Solomon codes.

Paris, April 1988 Gérard Cohen and Philippe Godlewski

LIST OF REFEREES

M. Beveraggi, J.C. Bic, B. Bouchon, P. Camion, G. Cohen, J.L. Dornstetter, P. Godlewski, C. Goutelard, A. Lobstein, P. Solé, R. Vallet, J. Wolfmann, G. Zemor

# COLLOQUE INTERNATIONAL

# "TROIS JOURNEES SUR LE CODAGE"

## PARRAINAGE SCIENTIFIQUE
### ADVISORY BOARD

J.-C. BERMOND
Directeur de Recherche CNRS
P. CAMION
Directeur de Recherche CNRS-INRIA
P. FAURRE
Directeur Général de la SAGEM
Membre de l'Institut
C. GUEGUEN
Professeur à l'ENST
Directeur de l'UA 820 du CNRS
P. LALLEMAND
Directeur scientifique de la DRET
D. LAZARD
Professeur à l'Université Paris VI
Directeur du Greco "Calcul formel"
D. LOMBARD
Directeur du CNET Paris B
D. PERRIN
Professeur à l'Université Paris VII, LITP
B. PICINBONO
Professeur à l'Université Paris-Sud, Orsay
Directeur du LSS, Gif sur Yvette

## COMITE D'ORGANISATION
### ORGANIZING COMMITTEE

C. GOUTELARD (LETTI)
L. RIZZI (DRET)
P. GODLEWSKI (ENST)
S. HARARI (GECT)

## COMITE DE PROGRAMME ET PUBLICATIONS
### PROGRAM COMMITTEE

G. COHEN (ENST)
P. GODLEWSKI (ENST)
C. GOUTELARD (LETTI)

## COMITE SCIENTIFIQUE
### SCIENTIFIC COMMITTEE

G. BATTAIL (ENST),
J. C. BIC (CNET)
B. BOUCHON (CNRS),
P. CAMION (CNRS),
M. CHARBIT (ENST, LSS),
P. CHARPIN (Paris VI),
G. COHEN (ENST),
J. CONAN (Polytec. de Montréal),
B. COURTEAU (Univer. de Sherbrooke),
J. L. DORNSTETTER (LCT),
Y. DRIENCOURT (Paris-VII),
P. GODLEWSKI (ENST),
C. GOUTELARD (LETTE),
D. HACCOUN (Polytec. de Montréal),
S. HARARI (GECT),
P. LAURENT (THOMSON-CSF DTC),
S. LEBEL (SCCST),
J. F. MICHON (Paris VII),
A. OISEL (BULL),
D. PERRIN (Paris-VII),
P. PIRET (PHILIPS R. L.),
A. POLI (AAECC),
J. WOLFMANN (GECT).

## PATRONAGE :
### SPONSORS :

- Groupe d'Etudes de Codage de la Direction des Recherches, Etudes et Techniques (DRET)
- Laboratoire d'Etudes des Transmissions Ionosphériques (LETTI)
- Bull S.A.
- Equipe "Traitement de l'information discrète" de l'ENS des Télécommunications
- Groupe d'Etudes du Codage de Toulon (GECT, Université de Toulon)
- Laboratoire d'Algèbre Appliquée et Codes Correcteurs (AAECC, Université Paul-Sabatier de Toulouse)
- Equipe "Théorie de l'information" du groupe de recherches "Claude-François Picard", CNRS
- Groupe d'études de Codage de l'Université Paris VII

# CONTENTS

# INFORMATION THEORY IN RANDOM FIELDS

Toby Berger

Department Systems at Connunications
CNRS UA 820
E.N.S.T
46, rue Barrault, 75634 Paris Cedex 13, France

School of Electrical Engineeering
and
Center for Applied Mathematics
Cornell University
Ithaca, New York 14853

*ABSTRACT*

A random field is a family of random variables with a multidimensional parameter set. Random fields provide mathematical models for distributed sources of information. Channels that link an input and an output random field also are of interest.

First, we describe in detail a celebrated result of random field theory to the effect that a random field has the Markov property if and only if it is a Gibbs state with a nearest neighbor potential. Next we lower bound the zero-error capacity of a certain binary random field channel by developing an efficient zero-error coding scheme. Finally, we consider algorithms for computing the stationary distribution of a time evolution mechanism. These algorithms, which have long been employed in mathematical statistical mechanics, also play a central role in simulated annealing.

## *1. A Basic Result of Markov Random Field Theory*

### *1.1 Binary Random Fields*

Let $\Lambda$ be a finite set. Each point $\lambda \in \Lambda$ is called a "site," and at site $\lambda$ we have a binary random variable $X_\lambda \in \{0,1\}$. Depending upon the application, the event $[X_\lambda = 1]$ may represent the presence (as opposed to the absence) of a particle at $\lambda$, a black (as opposed to a white) fascimile pixel at $\lambda$, a spin-up (as opposed to a spin-down) magnetic monent at $\lambda$, a radical (as opposed to a conservative) politician at $\lambda$, and so on. The collection $\{X_\lambda, \lambda \in \Lambda\}$ is called a random field.

### *1.2 Connectivity and Neighbors*

Interesting random fields reflect interactions between neighboring sites. To produce such models we impose a connectivity structure on $\Lambda$; that is, we define a graph $(\Lambda, e)$ where the edge set $e$ tells us which sites are neighbors. Let $c(x,y) = 1$ if there is an edge between sites $x$ and $y$ (i.e., if $x$ and $y$ are neighbors), and let $c(x,y) = 0$ otherwise.

Given any set of sites $A \subset \Lambda$, define the boundary of $A$ by the prescription

$$\partial A = \{y \notin A : c(x,y) = 1 \text{ for at least one } x \in A\}.$$

A simplex, or clique, is a set of sites any two distinct members of which are neighbors; formally, $B \subset \Lambda$ is a simplex if

$$(x, y \in B, x \neq y) \Rightarrow (c(x,y) = 1).$$

Note that for each $x \in \Lambda$, $\{x\}$ is a simplex.

### *1.3 Distributions and Potentials*

We associate with each $A \subset \Lambda$ a configuration of the random field, namely the one in which $X_\lambda = 1$ if $\lambda \in \Lambda$ and $X_\lambda = 0$ if $\lambda \notin A$. In the facsimile example, for instance, the configuration

$A$ is the picture that is black at each point in $A$ and white at each point in $A^c$. Let $\mu$ define a distribution over the space of possible configurations in the sense that $\mu(A)$ denotes the probability that the configuration is $A$. Then $\mu(A) \geq 0$ and

$$\sum_{A \subset \Lambda} \mu(A) = 1.$$

A distribution over configuration space is called a "state." It is often convenient to describe a state in terms of a so-called ***potential***, $V(\cdot)$, where $-\infty < V(A) < \infty$ for every $A \subset \Lambda$ and $V(\phi) = 0$. The Gibbs state $\pi(\cdot)$ with potential $V(\cdot)$ is defined by

$$\pi(A) = Z^{-1} \exp[V(A)], A \subset \Lambda$$

where the normalizing constant

$$Z = \sum_{A \subset \Lambda} \exp[V(A)]$$

is called the ***partition function.*** [In statistical mechanics applications, Z is a function of physical parameters, such as temperature, pressure, density, and field strength, that appear in the definition of $V(\cdot)$.] Note that every state $\mu$ for which $\mu(A) > 0$ for all $A \subset \Lambda$ is a Gibbs state with potential $V(A) = \log[\mu(A)/\mu(\phi)]$. The ***interaction potential*** corresponding to $V(\cdot)$ is the function

$$J(A) = \sum_{G \subset A} (-1)^{|A-G|} V(G)$$

where $|X|$ denotes the cardinality of $X$. (Since $J(\phi) = V(\phi) = 0$, $J(\cdot)$ is indeed a potential.) The principle of inclusion and exclusion readily yields the inversion

$$V(A) = \sum_{B \subset A} J(B).$$

Interesting mathematical models result when the potential is connected with the graph structure in a meaningful way. In particular, we say that $V(\cdot)$ is a nearest neighbor potential if the corresponding interaction potential $J(\cdot)$ is such that $J(B) \neq 0$ only if $B$ is a simplex of the graph.

### *1.4 Markov Random Fields*

#### *1.4.1 Configurations Plus or Minus Sites*

Let $A \subset \Lambda$ be a configuration, and let $\lambda \notin A$ be a site not in $A$. The configuration $A \cup \{\lambda\}$ corresponds to the facsimile picture in which $\lambda$ and all the sites in $A$ are colored black and the rest is left white. For compactness we write $A \cup \lambda$ rather than $A \cup \{\lambda\}$. Similarly, for $\lambda \in A$ let $A - \lambda$ denote the picture that results from starting with configuration $A$ but then painting $\lambda$ white. Moreover, define

$$A \oplus \lambda = \begin{cases} A \cup \lambda & \text{if } \lambda \notin A \\ A - \lambda & \text{if } \lambda \in A. \end{cases}$$

Also, for the boundary of the singleton $\{\lambda\}$ write $\partial\lambda$ rather than $\partial\{\lambda\}$.

#### *1.4.2 Conditional Probabilities*

Let $\mu(\cdot)$ define a distribution over configurations. Consider the conditional probability that the configuration on $\Lambda$ is $A \cup \lambda$, given that the subconfiguration on $\Lambda - \lambda$ is $A$. Since $A$ on $\Lambda - \lambda$ means either $A$ or $A \cup \lambda$ on $\Lambda$, we have

$$P(A \cup \lambda \quad \text{on} \quad \Lambda | A \quad \text{on} \quad \Lambda - \lambda) = \frac{\mu(A \cup \lambda)}{\mu(A) + \mu(A \cup \lambda)}.$$

The notation here is potentially confusing in that we expect the probability of an *intersection* in the numerator of a conditional probability but we appear to have the probability of a *union* there. However, $A \cup \lambda$ is indeed interpretable as the intersection of the two events $[A \text{ on } \Lambda - \lambda]$ and $[X_\lambda = 1]$. Thus, the preceding equation also can be expressed as

$$P(X_\lambda = 1 \,|\, A \,\text{on}\, \Lambda - \lambda) = \frac{\mu(A \cup \lambda)}{\mu(A) + \mu(A \cup \lambda)}.$$

Next, consider the conditional probability that the configuration contains $\lambda$ given that it is $A \cap \partial\lambda$ on $\partial\lambda$, in other words given that on the boundary of $\{\lambda\}$ the configuration agrees with the restriction of $A$ thereto. This conditioning allows the portion on the configuration external to $\lambda \cup \partial\lambda$ to be any subset whatsoever of $\Lambda - (\lambda \cup \partial\lambda)$. Thus

$$P(X_\lambda = 1 \,|\, A \cap \partial\lambda \,\text{on}\, \partial\lambda) = \frac{\sum\limits_{B \subset \Lambda - (\lambda \cup \partial\lambda)} \mu((A \cap \partial\lambda) \cup \lambda \cup B)}{\sum\limits_{B \subset \Lambda - (\lambda \cup \partial\lambda)} \mu((A \cap \partial\lambda) \cup B) + \mu((A \cap \partial\lambda) \cup \lambda \cup B)}$$

*1.4.3 Markovianness*

A random sequence $\{X_k, k = 0, \pm 1, \pm 2, \ldots\}$ is said to be Markovian if, given $X_s$ for any integer $s$, the families of random viariables $\{X_k, k < s\}$ and $\{X_k, k > s\}$ are conditionally independent. Expressed loosely, a random sequence is Markovian if, given the "present" $\{X_k, k = s\}$, the "past" $\{X_k, k < s\}$ becomes conditionally independent of the "future" $\{X_k, k > s\}$. The natural extension of this concept to random fields is that, for any set of sites $S \subset \Lambda$, given the "boundary" $\{X_\lambda, \lambda \in \partial S\}$, the "interior" $\{X_\lambda, \lambda \in S\}$ becomes conditionally independent of the "exterior" $\{X_\lambda, \lambda \notin S \cup \partial S\}$. It turns out that, if this property holds for every singleton $S = \{\lambda\}$, then it holds for any finite $S$. It follows that a binary random field $\{X_\lambda\}$ is Markovian if and only if for all $\lambda \in \Lambda$

$$P(X_\lambda = 1 \,|\, X_\nu, \nu \neq \lambda) = P(X_\lambda = 1 \,|\, X_\nu, \nu \in \partial\lambda)$$

Equivalently, $\{X_\lambda\}$ is Markovian if for all $\lambda \in A \subset \Lambda$,

$$P(X_\lambda = 1 \,|\, A \,\text{on}\, \Lambda - \lambda) = P(X_\lambda = 1 \,|\, A \cap \partial\lambda \,\text{on}\, \partial\lambda).$$

From the expressions derived in Section 1.4.2, we see that a distribution $\mu$ over configuration space is Markovian if, whenever $\lambda \notin A \subset \Lambda$, we have

$$\frac{\mu(A \cup \lambda)}{\mu(A) + \mu(A \cup \lambda)} = \frac{\sum\limits_{B \subset \Lambda - (\lambda \cup \partial\lambda)} \mu((A \cap \partial\lambda) \cup \lambda \cup B)}{\sum\limits_{B \subset \Lambda - (\lambda \cup \partial\lambda)} \mu((A \cap \partial\lambda) \cup B) + \mu((A \cap \partial\lambda) \cup \lambda \cup B)}$$

Formally $\{X_\lambda\}$ is the random field and $\mu(\cdot)$ is its density. However, it is common parlance to say that $\mu$ is a Markov random field if (i)$\mu(A) > 0$ for all $A \subset \Lambda$ and (ii) $\mu$ satisfies the preceding equation for all $\lambda \notin A \subset \Lambda$.

*1.5 The Fundamental Theorem*

A fundamental result of random field theory is the following:

*Theorem:* $\mu$ is a Markov random field if and only if it is a Gibbs state with a nearest neighbor potential.

The proof of this celebrated result involves the intermediary concept of a so-called nearest neighbor state. We say $\mu(\cdot)$ is a nearest neighbor state if $\mu(A) > 0$ for all $A \subset \Lambda$ and

$$\frac{\mu(A \cup \lambda)}{\mu(\lambda)} = \frac{\mu((A \cap \partial\lambda) \cup \lambda)}{\mu(A \cap \partial\lambda)}, \lambda \notin A \subset \Lambda.$$

A series of elegant steps, which may be found in Preston [1], establish that $\mu$ is a Markov random field if and only if it is a nearest neighbor state, and then that $\mu$ is nearest neighbor state if and only if it is a Gibbs state with a nearest neighbor potential. In the course of the proof there also arises the following interesting characterization of a nearest neighbor potential: $V$ is a nearest neighbor potential if and only if for all $x, y \in \Lambda$, $x \neq y$, such that $c(x, y) = 0$ and for all $G \subset \Lambda - x - y$,

$$V(G \cup x \cup y) - V(G \cup x) - V(G \cup y) + V(G) = 0.$$

(To see the connection between this condition and the preceding characterizations of Markovianness, add $V(G \cup y) - V(G)$ to both sides, exponentiate, and then compare the result with the definition of a nearest neighbor state.)

*1.6 Decompositoin of Binary Random Fields*

Consider a binary Markov random field whose neighbor structure is specified by a countable graph with nodes of uniformly bounded degree. In joint work with Bruce Hajek [2], we have shown that any such Markov random field can be represented as the nodewise modulo 2 sum of two independent binary random fields one of which is white binary noise of positive weight. This permits us to evaluate the field's rate-distortion function exactly over an interval of small distortions.

## *2 Zero-Error Capacity of a 2-D Binary Channel*

*2.1 Problem Formulation*

Consider the integer lattice, $\Lambda = \{(i, j) : i, j \quad \text{integers}\}$, and the partition

$$\Lambda = D_1 \cup D_2$$

with $D_1 = \{(i, j) \in \Lambda : i+j \quad \text{odd}\}$ the set of the odd diagonals and $D_2 = \{(i, j) \in \Lambda : i+j \quad \text{even}\}$ the set of the even diagonals of $\Lambda$. Let

$$\underline{x} = \{x_t\}_{t \in D_1} \qquad \text{and} \qquad \underline{y} = \{y_t\}_{t \in D_2}$$

be two families of random variables with bi-dimensional index, i.e., two random fields.

We consider $\underline{x}$ and $\underline{y}$ as the input and output field, respectively, for a 2-D channel system $C_\Lambda$, described as follows: let the set of neighbors of $t \in \Lambda$ be

$$N(t) = \{t + (0, 1), t + (0, -1), t + (1, 0), t + (-1, 0)\},$$

and let $\underline{y}_T$ and $\underline{x}_{N(t)}$ be the restrictions of $\underline{y}$ to $T \subset D_2$ and of $\underline{x}$ to $N(t)$, respectively. Define

$$C_\Lambda = [\{0, 1\}^{D_1}, \nu(\underline{y}|\underline{x}), \{0, 1\}^{D_2}],$$

where $\{0, 1\}^{D_1}$ and $\{0, 1\}^{D_2}$ are the spaces in which $\underline{x}$ and $\underline{y}$, respectively, take values, and $\nu$ is the channel transition measure defined by:

$$\nu(\underline{y}_T|\underline{x}) = \prod_{t \in T} q(y_t|\underline{x}_{N(t)})$$

for every finite subset $T \subset D_2$; here, $q$ is the conditional probability of the elementary channel for the channel system $C_\Lambda$ specified as follows via Figure 1. First, we group the input words $\tilde{x} = (x_1, x_2, x_3, x_4)$ into the subsets $\{A_j\}_{j=0,\ldots 4}$ defined by:

$$A_i = \{\tilde{\underline{x}} : \sum_{j=1}^{4} x_j = i\}.$$

Let

$$\epsilon = e^{-\beta J}$$

where $\beta = \frac{1}{kT}$, with $k$ a physical constant, $T$ the absolute temperature, and $J$ a constant determining the strength of the interaction. Then Figure 1 shows the conditional probability $q$ for the elementary channel under consideration.

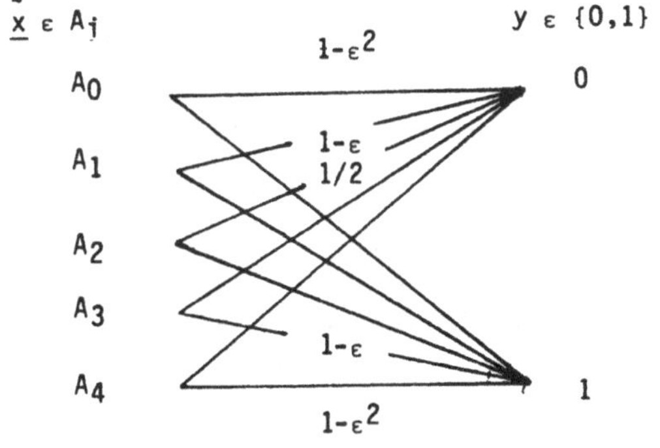


Figure 1. Line diagram representation of $q(y_t|\underline{x}_{N(t)})$.

The case $T = 0$ is shown in Figure 2 where:

$$M_0 = A_0 \cup A_1 = \{\tilde{\underline{x}} : \sum_{j=1}^{4} x_j < 2\}$$

$$M_1 = A_3 \cup A_4 = \{\tilde{\underline{x}} : \sum_{j=1}^{4} x_j > 2\}$$

$$M_2 = A_2 = \{\tilde{\underline{x}} : \sum_{j=1}^{4} x_j = 2\}.$$

The channel in Figure 2 can be related to a *voter system* where a majority of ones or zeros at the input imposes a one or a zero with probability one at the output. Total uncertainty remains when the inputs is comprised of two ones and two zeros.

*2.2 Channel System Zero-Error Capacity*

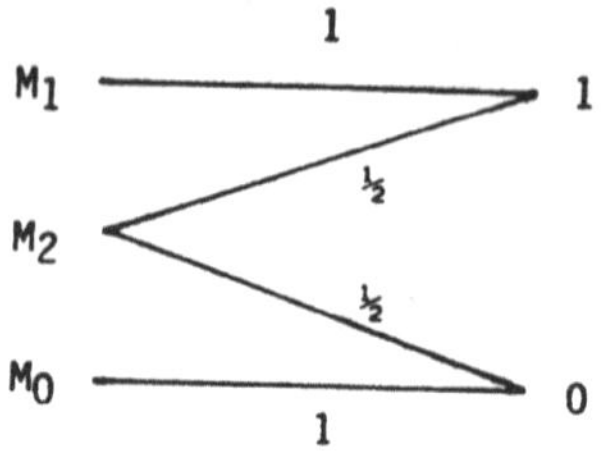


Figure 2. Elementary channel for $T = 0$.

Let us consider a discrete finite channel system, i.e., a system with a finite number of input and output sites. We will say that two input patterns are ***adjacent*** if there exists an output pattern that can be caused with non-zero probability by either of them. Adjacency of input patterns can be described using graphs as in Figure 3.

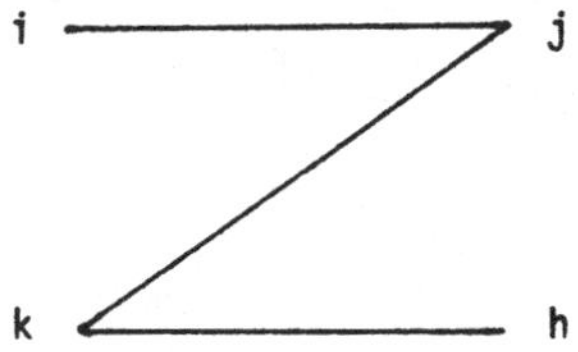


Figure 3. Adjacency diagram.

An input pattern (node) $i$ is connected to an output pattern (node) $j$ when the channel probability $p(j|i)$ is not zero. Two input patterns are adjacent if they are connected to a common output node.

$$A_{ij} = \begin{cases} 1 & \text{if input } i \text{ is adjacent to output } j \text{ or } i = j \\ 0 & \text{otherwise} \end{cases}$$

The least upper bound of all rates at which it is possible to transmit at a positive rate with zero probability of error is called the ***zero-error capacity*** of the channel system. If we let $M_0(n)$ be the largest number of patterns a code on $n$ input locations can have such that no two of them are adjacent, then the zero-error capacity is

$$C_0(n) = \frac{1}{n} \log M_0(n).$$

In what follows we will consider the zero-error capacity for infinite channel systems, namely

$$C_0 = \limsup_{n \to \infty} C_0(n)$$

for a suitable sequence of finite channel systems with zero-error capacities $C_0(n)$.

### *2.3 Zero-Error Capacity of Basic Finite Channel Systems*

Let us consider the 2-D channel system introduced in Section 2.1. Assume $T = 0$, or equivalently, $\epsilon = 0$. Our goal is to determine the zero-error capacity for this infinite system, or at least to obtain meaningful lower and upper bounds for it. To gain insight into the basic mechanisms involved, it is useful to start from the analysis of some fundamental finite subsystems of the infinite system.

We will consider the following channel systems:

(a) The basic channel element (Figure 4)
(b) The $X_0$ system (Figure 5)
(c) The $S_1$ system (Figure 6).

```
x   x
  o
x   x
```

Figure 4. The basic channel element.

```
x   x   x
  o   o
x   x   x
```

Figure 5. The $S_0$ system.

```
x   x   x   x
  o   o   o
x   x   x   x
  o   o   o
x   x   x   x
```

Figure 6. The $S_1$ system.

*2.3.1 The Basic Channel Element.*

In this case the concept of capacity makes sense only when one considers having infinitely many independent repetitions of the finite-input channel. The two non-adjacent input patterns for the basic channel element can be obtained simply by assigning three out of the four input bits either the value 0 or the value 1. The fourth location can have an arbitrary value. The zero-order capacity, obtained by normalizing with respect to the number of input locations, is thus

$$C_0 = (\log 2)/4 = 1/4.$$

*2.3.2 The $S_0$ Channel System*

In this case all the non-adjacent input patterns can be obtained by grouping the input bits as in Figure 7 and assigning a value 0 or 1 to all the members of each group.

The number of non-adjacent patterns so determined is optimal and consequently

$$C_0^0 = (\log 4)/6 = 1/3.$$

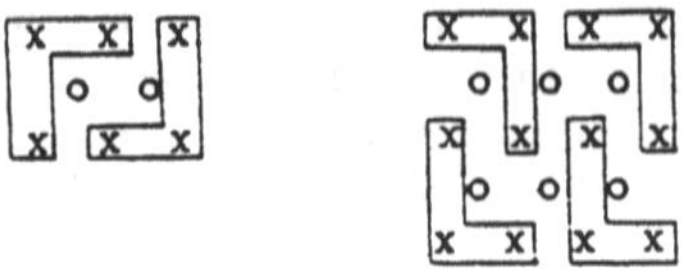

Figure 7. Grouping of inputs for $S_0$ and $S_1$ systems.

*2.3.3 The $S_1$ Channel System*

If we grouped the input locations as in Figure 7 and assigned zeros and ones to all the elements of each group, we would obtain a set of $2^4 = 16$ non-adjacent patterns and a rate $R = 4/12 = 1/3$. Note that the output bits marked in black in Figure 7 would be irrelevant for discriminating among these input patterns. It is possible, however, to exploit the marked ouptut bits so as to determine up to 32 non-adjacent input patterns as shown in Table 1. The rate for this code is then

$$R(S_1) = R_1 = (\log 32)/12 = 0.41666.$$

Is $R_1$ the maximum achievable rate for zero-order codes on $S_1$? In this regard we note that a maximum of $2^6 = 64$ input patterns can be non-adjacent ($S_1$ has 6 output bits). This implies that

$$C_0^1 \leq (\log 64)/12 = 0.5.$$

Thus

$$0.4166 \leq C_0^1 \leq 0.5.$$

Looking at Table 1 we observe that 28 out of the 32 input patterns listed induce a single output pattern, while the remaining 4 have two possible images. It is easy to verify that each input pattern not listed in Table 1 is consistent with at least one ouptut pattern listed there. That is, every non-listed input pattern is adjacent to at least one of the listed input patterns. The zero-error code in Table 1 is therefore at least a local maximum on $S_1$, if not the global maximum.

*2.4 Achievable Rates for Infinite Channel Systems*

We now consider infinite subsystems of the 2-D channel system. The computation of achievable rates for such subsystems provides lower bounds to the zero-error capacity and illuminates some of the basic mechanisms involved in the limiting process. Two expansion schemes will be analyzed, stripe expansion and slanted expansion, in both of which a new $S_1$ block is added at each step together with the necessary extra output locations.

*2.4.1 Stripe Expansion*

The stripe expanding system is shown in Figure 8.

The two output locations between abutting $S_1$ blocks often can be used to provide information for descriminating among input patterns. Consider, for example, the juxtaposition of patterns (1) and (22) in Table 1. By modifying the values at some of the input locations on the interface, we can obtain 4 non-adjacent input patterns by forcing either 11, 10, 01, or 00 at the two output locations between blocks (see Figure 9).

The output patterns within the $S_1$ blocks themselves remain unchanged. It is clear that what is relevant about an input pattern in this regard is the degree to which its rightmost column and/or its leftmost column can be modified without changing the output pattern within the $S_1$ block.

| # | Input | Output | Label I | # | Input | Output | Label I |
|---|---|---|---|---|---|---|---|
| | 0 0 0 0 | 0 0 0 | 2 | | 0 0 1 1 | 0 1 1 | 0 |
| 1 | 0 0 0 0 | 0 0 0 | | 17 | 0 1 1 1 | 0 1 1 | |
| | 0 0 0 0 | | 2 | | 0 0 1 1 | | 0 |
| | 0 0 0 0 | 0 0 0 | 1 | | 0 0 1 1 | 0 0 1 | |
| 2 | 1 0 0 0 | 1 0 0 | | 18 | 0 0 0 1 | 0 0 1 | |
| | 1 1 0 0 | | 0 | | 0 0 1 1 | | 1 |
| | 1 1 0 0 | 1 0 0 | 0 | | 0 0 1 1 | 0 0 1 | 0 |
| 3 | 1 0 0 0 | 0 0 0 | | 19 | 1 0 1 1 | 1 1 1 | |
| | 0 0 0 0 | | 1 | | 1 1 1 1 | | 1 |
| | 1 1 1 0 | 1 1 0 | 0 | | 1 1 1 1 | 1 1 1 | 1 |
| 4 | 0 1 0 0 | 0 0 0 | | 20 | 0 1 1 1 | 1 1 1 | |
| | 0 0 0 0 | | 1 | | 1 1 1 1 | | 1 |
| | 1 1 0 0 | 1 0 0 | 0 | | 1 1 1 1 | 1 1 1 | 1 |
| 5 | 1 0 0 0 | 1 0 0 | | 21 | 0 1 1 1 | 0 1 1 | |
| | 1 1 0 0 | | 0 | | 0 0 1 1 | | 2 |
| | 1 1 0 0 | 1 1 0 | 0 | | 1 1 1 1 | 1 1 1 | 2 |
| 6 | 1 1 1 0 | 1 1 0 | | 22 | 1 1 1 1 | 1 1 1 | |
| | 1 1 0 0 | | 0 | | 0 0 1 1 | | 2 |
| | 0 0 0 0 | 0 0 0 | 1 | | 1 1 0 0 | 1 0 0 | 0 |
| 7 | 0 0 0 1 | 0 0 1 | | 23 | 1 0 0 1 | 0 0 1 | |
| | 0 0 1 1 | | 0 | | 0 0 1 1 | | 0 |
| | 0 0 0 0 | 0 0 0 | 1 | | 0 0 1 1 | 0 1 1 | 0 |
| 8 | 0 0 1 0 | 0 1 1 | | 24 | 0 1 1 1 | 1 1 0 | |
| | 0 1 1 1 | | 0 | | 1 1 0 0 | | 0 |
| | 0 0 0 0 | 0 0 0 | 0 | | 1 1 1 0 | 1 1 0 | 1 |
| 9 | 1 0 1 0 | 1 1 1 | | 25 | 1 1 0 0 | 1 0 0 | |
| | 1 1 1 1 | | 0 | | 1 1 1 1 | | 1 |
| | 1 1 0 0 | 1 1 0 | 0 | | 1 0 0 0 | 1 0 0 | 1 |
| 10 | 0 1 1 0 | 0 1 1 | | 26 | 1 1 0 0 | 1 1 0 | |
| | 0 0 1 1 | | 0 | | 1 1 1 0 | | 1 |
| | 1 1 0 0 | 1 1 0 | 0 | | 0 1 1 1 | 0 1 1 | 0 |
| 11 | 1 1 1 0 | 1 1 1 | | 37 | 0 0 1 1 | 0 0 1 | |
| | 1 1 1 1 | | 1 | | 0 0 0 1 | | 0 |
| | 0 0 1 1 | 0 0 1 | 0 | | 0 0 0 1 | 0 0 1 | 1 |
| 12 | 0 0 0 1 | 0 0 0 | | 28 | 0 0 1 1 | 0 1 1 | |
| | 0 0 0 0 | | 1 | | 0 1 1 1 | | 1 |
| | 0 0 1 1 | 0 0 1 | 0 | | 0 0 0 0 | 0 0 0 | 0 |
| 13 | 1 0 0 1 | 1 0 0 | | 29 | 0 1 0 0 | X 0 1 | |
| | 1 1 0 0 | | 0 | | 0 1 1 0 | | 0 |
| | 1 1 1 1 | 1 1 1 | 0 | | 1 1 1 1 | 1 1 1 | 1 |
| 14 | 0 1 0 1 | 0 0 0 | | 30 | 1 0 1 1 | X 0 1 | |
| | 0 0 0 0 | | 0 | | 1 0 0 1 | | 0 |
| | 1 1 1 1 | 1 1 1 | 1 | | 0 1 1 0 | 0 1 X | 0 |
| 15 | 1 1 0 1 | 1 0 0 | | 31 | 0 0 1 0 | 0 0 0 | |
| | 1 0 0 0 | | 0 | | 0 0 0 0 | | 0 |
| | 1 1 1 1 | 1 1 1 | 1 | | 1 0 0 1 | 1 0 X | 0 |
| 16 | 1 1 1 0 | 1 1 0 | | 32 | 1 1 0 1 | 1 1 1 | |
| | 1 1 0 0 | | 0 | | 1 1 1 1 | | 0 |

Table 1: Zero-error codebook for channel system $S_1$.

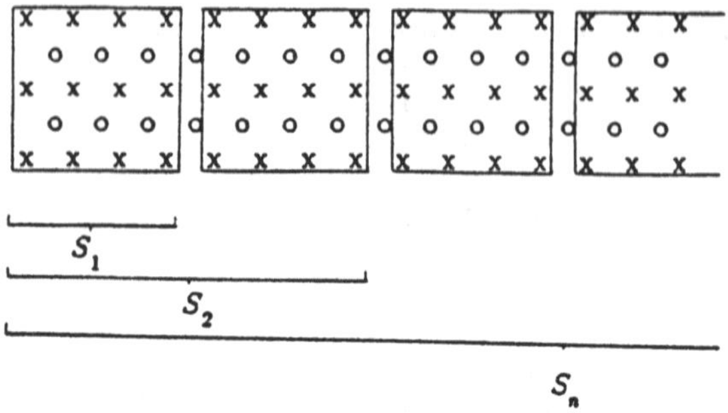


Figure 8. The stripe expanding system.
($x$ = input location, $o$ = output location)

| Input pattern | Output pattern |
|---|---|
| 0 0 0 1 1 1 1 1 | |
| | 0 0 0 1 1 1 1 |
| 0 0 0 0 1 1 1 1 | |
| | 0 0 0 1 1 1 1 |
| 0 0 0 1 1 1 1 1 | |
| 0 0 0 1 1 1 1 1 | |
| | 0 0 0 1 1 1 1 |
| 0 0 0 0 1 1 1 1 | |
| | 0 0 0 0 1 1 1 |
| 0 0 0 0 0 1 1 1 | |
| 0 0 0 0 0 1 1 1 | |
| | 0 0 0 0 1 1 1 |
| 0 0 0 0 1 1 1 1 | |
| | 0 0 0 1 1 1 1 |
| 0 0 0 1 1 1 1 1 | |
| 0 0 0 0 0 1 1 1 | |
| | 0 0 0 0 1 1 1 |
| 0 0 0 0 1 1 1 1 | |
| | 0 0 0 0 1 1 1 |
| 0 0 0 0 0 1 1 1 | |

Figure 9. Some codewords on $S_2$

In order to characterize the patterns in Table 1 in this sense, we introduce additional labels as in Table 2 (Label II). The symbols used in Table 2 relate to the number of input bits that can change value without affecting the output pattern as follows:

0 — No input bit can be modified
2 — Two input bits can be modified

| Pattern # | Label II Left | Label II right | Pattern # | Label II Left | Label II Right |
|---|---|---|---|---|---|
| 1 | 2 | 2 | 17 | 0 | 2 |
| 2 | 0 | 2 | 18 | 2 | 0 |
| 3 | 0 | 2 | 19 | 0 | 1L |
| 4 | 0 | 1L | 20 | 0 | 2 |
| 5 | 0 | 2 | 21 | 0 | 2 |
| 6 | 2 | 0 | 22 | 2 | 2 |
| 7 | 2 | 0 | 23 | 0 | 0 |
| 8 | 1U | 0 | 24 | 0 | 0 |
| 9 | 0 | 0 | 25 | 1U | 1L |
| 10 | 0 | 0 | 26 | 1L | 1U |
| 11 | 2 | 0 | 27 | 1L | 1U |
| 12 | 2 | 0 | 28 | 1U | 1L |
| 13 | 0 | 0 | 29 | 1L | 1U |
| 14 | 0 | 0 | 30 | 1L | 1U |
| 15 | 1U | 0 | 31 | 1L | 1U |
| 16 | 2 | 0 | 32 | 1L | 1U |

Table 2. Label II for patterns in Table 1

1U — The bit on the upper row only can be modified

1L — The bit on the lower row only can be modified

In the label column the position of the entry, right or left, refers to the rightmost or the leftmost column of the $S_1$ block in question. Note that we have used the output bits on the interface for purposes of discrimination only for patterns with the columns on the interface having different values and whenever on both sides it is possible to modify the inputs at the same level, $U$ or $L$. (It can be shown that this engenders no less of generality.) Using these considerations and Table 2, we proceed to determine the number of non-adjacent input patterns for a generic system $S_k$.

Consider first the case $k = 2$. In Table 3 we give the number of non-adjacent patterns on $S_2$ whose sixteen leftmost input bits show the indicated $S_1$ pattern number

The total number of non-adjacent configurations obtained for $S_2$ in this way is

$$N(S_2) = N(2) = 1224,$$

and the corresponding rate is

$$R(S_2) = (\log 1224)/24 = 0.42739.$$

To obtain a recursive equation for $N(k)$, note that the input pattern on the last two columns of $S_{k-1}$ block conditions the number of non-adjacent configurations that can be obtained by adding another $S_1$ block as shown in Table 4.

| Pattern # | # of patterns on $S_2$ | Pattern # | # of patterns on $S_2$ |
|---|---|---|---|
| 1 | 49 | 17 | 49 |
| 2 | 49 | 18 | 32 |
| 3 | 49 | 19 | 39 |
| 4 | 39 | 20 | 49 |
| 5 | 49 | 21 | 49 |
| 6 | 32 | 22 | 49 |
| 7 | 32 | 23 | 32 |
| 8 | 32 | 24 | 32 |
| 9 | 32 | 25 | 39 |
| 10 | 32 | 26 | 38 |
| 11 | 32 | 27 | 38 |
| 12 | 32 | 28 | 39 |
| 13 | 32 | 29 | 38 |
| 14 | 32 | 30 | 38 |
| 15 | 32 | 31 | 38 |
| 16 | 32 | 32 | 38 |

Table 3

| Class | Pattern on last two columns | # of non-adjacent patterns obtainable |
|---|---|---|
| A | 0 0 1 1<br>0 0 1 1<br>0 0 1 1 | 49 |
| B | 1 0 0 1<br>0 0 1 1<br>0 0 1 1 | 39 |
| C | 0 0 1 1 0 1 1 0<br>0 0 1 1 0 1 1 0<br>1 0 0 1 1 1 0 0 | 38 |
| D | Any other | 32 |

Table 4

| Pattern Class on $S_{k-1}$ | Distribution of new patterns on $S_k$ A | B | C | D | Total |
|---|---|---|---|---|---|
| A | 11 | 5 | 9 | 24 | 49 |
| B | 9 | 4 | 9 | 14 | 39 |
| C | 9 | 5 | 6 | 18 | 38 |
| D | 8 | 4 | 6 | 14 | 32 |

Table 5

To determine the coefficients in the system of recursive equations for $N(k)$, we need to know, for each of the four classes in Table 4, the distribution of the new patterns over these four classes. That information is in Table 5.

Let $Z$ be the transpose of the matrix in Table 5. Let $\underline{\mathrm{N}}(n)$ be the column vector

$$\mathrm{N}(n) = (N_A(n), N_B(n), N_C(n), N_D(n))^T$$

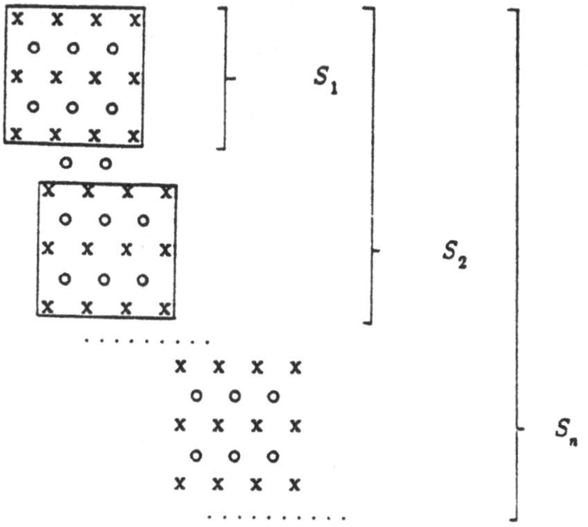


Figure 10. The slanted expanding system.

where $N_1(n)$ is the number of non-adjacent patterns on the stripe expanding system $S_n$ belonging to class $I, I = A, B, C,$ or $D$. Then

$$\underline{N}(n+1) = Z\underline{N}(n) \quad \text{for every} \quad n \geq 1.$$

and

$$N(n+1) = (1,1,1,1)\underline{N}(n+1).$$

By the Perron-Frobenius Theorem, the maximum eigenvalue $\lambda_{\max}$ of $Z$ is real and positive. We computed $\lambda_{\max} = 37.9678$. In this case it is possible to use the following asymptotic approximation

$$R(S_\infty) = \lim_{n\to\infty} R(S_n) = \lim_{n\to\infty} (\log N(n))/N_I(n)$$

$$= \lim_{n\to\infty} (\log \lambda_{\max})^n/12n = (\log 37.9678)/12 = 0.43723,$$

where $N_I(n) = 12n$ is the number of input points for $X_n$.

*2.4.2 Slanted Expansion.*

The slanted expanding system is shown in Figure 10. To exemplify the mechanisms involved, we consider again the juxtaposition of patterns (1) and (22) of Table 1. Figure 11 shows the 4 non-adjacent patterns on $S_2'$ obtained by modifying input bits on the interface.

Again it is important to determine which patterns can be modified on their first and last rows without affecting the output configuration. Using an approach parallel to the one described in Section 2.4.1 we determine that the asymptotic zero-error rate of the slanted expansion scheme is 0.43221, which is not as high as that of the strip expansion.

*2.5 Lower and Upper Bounds to the Zero-Error 2-D Capacity*

Consider a finite or semi-infinite subsystem $S$ of the 2-D channel system introduced above. Assume that we can tile the infinite lattice with a collection of $S$ systems as, for instance, in Figure 12. Here the output bit locations between blocks are represented by small circles.

| Input | output | Input | Output |
|---|---|---|---|
| 0 0 0 0 | | 0 0 0 0 | |
| | 0 0 0 | | 0 0 0 |
| 0 0 0 0 | | 0 0 0 0 | |
| | 0 0 0 | | 0 0 0 |
| 0 1 0 0 | | 0 0 0 1 | |
| | 1 0 | | 0 1 |
| 1 1 0 1 | | 0 1 1 1 | |
| | 1 1 1 | | 1 1 1 |
| 1 1 1 1 | | 1 1 1 1 | |
| | 1 1 1 | | 1 1 1 |
| 1 1 1 1 | | 1 1 1 1 | |
| 0 0 0 0 | | 0 0 0 0 | |
| | 0 0 0 | | 0 0 0 |
| 0 0 0 0 | | 0 0 0 0 | |
| | 0 0 0 | | 0 0 0 |
| 0 0 0 0 | | 0 1 0 1 | |
| | 0 0 | | 1 1 |
| 0 1 0 1 | | 1 1 1 1 | |
| | 1 1 1 | | 1 1 1 |
| 1 1 1 1 | | 1 1 1 1 | |
| | 1 1 1 | | 1 1 1 |
| 1 1 1 1 | | 1 1 1 1 | |

Figure 11. Some codewords on $S_2'$.

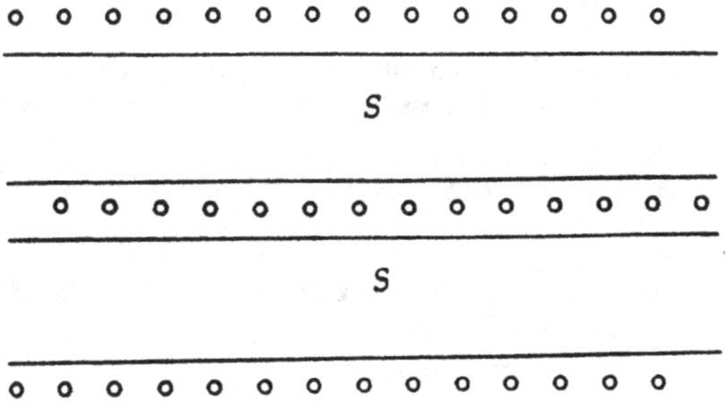


Figure 12. Tiling of the infinite lattice.

To obtain a lower bound to the zero-order capacity of the global system, we can disregard the information possibly carried by the output bits between blocks. It is thus clear that, if $R_s$ is a zero-error achievable rate for $S$, then the infinite system capacity $C_0$ must satisfy

$$R_S \leq C_0$$

The best lower bound we have obtained in this way is

$$C_0^{2-D} \geq 0.43723$$

by using the result of the strip expansion problem.

It is relevant at this point to discuss briefly the relationship between the ratio of the numbers of input and output locations and the zero-error capacity of a channel system. Since each output location carries at most one bit of information, we have that

$$C_0(S) \leq \frac{N_0(S)}{N_1(S)}$$

where $C_0(S)$ is the zero-error capacity for a generic channel system $S$ and $N_0(S)$ and $N_1(S)$ are the numbers of output and input locations of $S_k$ respectively. As an example consider the infinite strip channel system $S'_\infty$ analyzed above. If $S'_k$ is the finite strip system with $k$ blocks, we have that

$$N_I(S'_k) = 12k$$

$$N_0(S'_k) = 6k + 2(k-1)$$

and

$$C_0(S'_\infty) \leq 0.67.$$

This crude upper bound provides a yardstick for assessing the lower bounding techniques discussed earlier in this section.

## *3. Gibbs State Metropolis Algorithm and Simulated Annealing*

### *3.1. Gibbs States*

Gibbs states are used in statistical physics as models for steady-state distributions over the possible configurations of a physical system in dynamic equilibrium. Often the system is modeled by "particles" that interact in either discrete or continuous time. On a sufficiently microscopic time scale, such time evolutions proceed through a sequence of configurations each of which differs from its predecessor and its successor solely by virtue of a change in $X_\lambda$ at one particular $\lambda$. Under appropriate assumptions the corresponding sequence of configurations constitutes a Markov process. The Metropolis algorithm described in Section 3.2 shows how to choose the transition probabilities of this Markov process in such a way that a prescribed Gibbs state $\pi$ will be the equilibrium distribution of the Markov process. In the event that $\pi$ has a nearest neighbor potential (and hence is a Markov random field), we shall see that the Metropolis iterations become computationally trivial. Some of the many consequences of this computational simplification are discussed in Section 3.3

### *3.2. Metropolis Algorithm*

Let $\pi$ be a Gibbs state on $\{0,1\}$ $\Lambda$ with potential $V$. We proceed to construct a Markov chain on configuration space (i.e., on $\{0,1\}$ $\Lambda$) that has $\pi$ as its unique equilibrium distribution. (This also can be done in continuous time by means of an appropriate matrix of transition rates, but we shall treat only the discrete time case.)

Let $p$ be any strictly positive probability distribution over $\Lambda$. That is, $p : \Lambda \to (0,1)$ and $\sum_\lambda p(\lambda) = 1$. To construct our Markov chain, we must exhibit its transition probabilities, $Q(B|A)$, where $A$, $B \subset \lambda$. We do this by means of conditions (i) and (ii) below.

$$(i) \quad Q(B|A) = 0 \quad \text{unless either} \quad B = A \quad \text{or} \quad B = A \oplus \lambda \quad \text{for some} \quad \lambda \in \Lambda.$$

$$(ii) \quad Q(A \oplus \lambda|A) = p(\lambda)\max[1, \pi(A \oplus \lambda)/\pi(A)]$$

Condition (i) says that the state always either remains unchanged or transits to a state in which one and only one of the $X_\lambda$ is complemented. Condition (ii) says that in order to go to $A \oplus \lambda$ from $A$, one first must pick this $\lambda$ via random selection according to $p(\cdot)$; then, if $\pi(A \oplus \lambda) \geq \pi(A)$, the next state will be $A \oplus \lambda$, whereas if $\pi(A \oplus \lambda) < \pi(A)$, the state will transit to $A \oplus \lambda$ with probability $\pi(A \oplus \lambda)/\pi(A)$ and will stay at $A$ with probability $1 - \pi(A \oplus \lambda)/\pi(A)$. It follows that

$$Q(A|A) = \sum_{\{\lambda:\pi(A\oplus\lambda)<\pi(A)\}} p(\lambda)[1 - \pi(A \oplus \lambda)/\pi(A)].$$

*Theorem.* $\pi$ is the unique equilibrium distribution of the Markov chain whose transition matrix is the above-specified $Q(\cdot|\cdot)$.

*Proof.* It is clear that any configuration can be reached from any other in $|\Lambda|$ steps, so $Q^{|\Lambda|} > 0$. It follows from Markov's theorem that $Q$ has a unique equilibrium distribution (i.e., a limiting distribution). It therefore remains only to show that $\pi$ is a equilibrium distribution of $Q$. With reference to the partial state transition matrix shown below, we see that

$$(\pi Q)(B) = \sum_{\lambda:\pi(B\oplus\lambda)\le\pi(B)} \pi(B \oplus \lambda)p(\lambda) + \sum_{\lambda:\pi(B\oplus\lambda)>\pi(B)} \pi(B \oplus \lambda)p(\lambda)\frac{\pi(B)}{\pi(B \oplus \lambda)}$$

$$+ \sum_{\lambda:\pi(B\oplus\lambda)<\pi(B)} \pi(B)p(\lambda)\left[1 - \frac{\pi(B \oplus \lambda)}{\pi(B)}\right]$$

The first term represents the probability of coming to $B$ from some less $\pi$-probable state, the second the probability of coming to $B$ from some more $\pi$-probable state, and the third the probability of being in $B$ already and staying there. Since the third terms in unaffected by including any $\lambda$ such that $\pi(B \oplus \lambda) = \pi(B)$, algebra yields

$$(\pi Q)(B) = \sum_{\lambda:\pi(B\oplus\lambda)\le\pi(B)} \pi(B)p(\lambda) + \sum_{\lambda:\pi(B\oplus\lambda)>\pi(B)} \pi(B)P(\lambda)$$

$$= \pi(B)\sum_{\lambda\in\Lambda} p(\lambda) = \pi(B). \quad \text{Q.E.D.}$$

The so-called Metropolis algorithm consists of picking some initial configuration $A_0$ and then generating a specific sequence of successive states $A_1, A_2, \ldots$ according to the rule

$$Pr(A_1, A_2, \ldots, A_n) = \prod_{k=1}^{n} Q(A_k|A_{k-1}).$$

We discuss in Section 3.3 why one would care to generate such a sequence of configurations. For now let us focus on the computational buden inherent in the $k^{\text{th}}$ Metropolis update, i.e., the one that converts $A_{k-1}$ into $A_k$. It involves

(i) The random choice of a $\lambda \in \Lambda$ accroding to $p(\cdot)$

(ii) The calculation of

$$r_\lambda = \pi(A_{k-1} \oplus \lambda)/\pi(A_{k-1}).$$

(iii) The generation of a Bernoulli r.v. with probabilities $(r_\lambda, 1 - r_\lambda)$ in the event that $r_\lambda < 1$.

Step (ii) simpilifies significantly if $\pi$ is a nearest neighbor state, since in that event

$$\begin{aligned} r_\lambda = & \exp[V(A_{k-1} \oplus \lambda) - V(A_{k-1})] \\ = & \exp[\sum_{B \subset A_{k-1} \oplus \lambda} J(B) - \sum_{B \subset A_{k-1}} J(B)] \\ = & \exp[\sum_{\substack{\text{cliques } B \text{ in } A_{k-1} \oplus \lambda \\ \text{but not in } A_{k-1}}} J(B) - \sum_{\substack{\text{cliques in } A_{k-1} \\ \text{but not in } A_{k-1} \oplus \lambda}} J(B)] \\ = & \begin{cases} \exp[\sum_{\substack{\text{cliques } B \text{ in} A_{k-1} \oplus \lambda \\ \text{involving } \lambda}} J(B) & \text{if } \lambda \notin A_{k-1} \\ \exp[-\sum_{\substack{\text{cliques } B \text{ in } A_{k-1} \\ \text{involving } \lambda}} J(B) & \text{if } \lambda \in A_{k-1} \end{cases} \end{aligned}$$

This says that $r_\lambda$ can be calculated by a simple computation confined entirely to the neighborhood of $\lambda$. One merely calculates the $J$-values of the cliques created (or destroyed) when $\lambda$ is added (or deleted). Moreover, step (iii) also is greatly simplified in many instances in which $\pi$ is a nearest neighbor state, because it is often true that there is a $J$-preserving isomorphism between the sets of cliques to which any $\lambda$ belong and those to which any other site $\lambda'$ belongs. In that event the set of possible values of $r_\lambda$ does not depend on $\lambda$, so the Bernoulli r.v. can be generated off-line *a priori*. In some problems, moreover, it is possible to do Metropolis updates in a massively parallel fashion at many deterministically selected sites at once rather than to have to choose individual sites at random and update them.

*3.3. Model Verification and Simulated Annealing*

Suppose one wishes to assertain whether or not a particular potential $V(\cdot)$ accurately models the equilibrium statistical behavior of some system of interacting "particles". Although in principle one could compute $\pi(A) = Z^{-1}\exp(V(A))$ for each $A$, this usually proves infeasible in practice, since there are $2^{|\Lambda|}$ choices of $A$ even for a binary random field, and $|\Lambda|$ generally is in the thousands or in the millions even in small laboratory experiments or computer simulations. Instead, one uses the Metropolis algorithm to generate a configuration $A_n$ for some large $n$. The theorem in Section 3.2 assures us that as $n \to \infty$, $A_n$ has high probability of being a "typical" configuration in the sense that sample macroscopic quantities computed from it (energy, entropy, specific heat, magnetization, pressure, boson populations, etc.) should be close to thier respective mean values over all configurations calculated with respect to $\pi$. Therefore, one compares these sample quantities for $A_n$ with their experimentally determined counterparts and declares the model to be satisfactory if there is good agreement.

Often $V$ is of the form

$$V(A) = \frac{1}{kT} V^*(A)$$

where $k$ is Boltzmann's constant and $T$ is absolute temperature. As $T \to 0$ for fixed $V^*$, $\pi(A)$ becomes concentrated degenerately at the value of $A$ that maximizes $V^*(A)$ if there is a unique such $A$. Some optimization theorists use this fact in a technique they call simulated annealing.

They wish to maximize $V^*$, where $V^*$ depends on a large number of variables and does not seem to be amenable to calculus-based analysis. In simulated annealing a sequence of temperatures $T_1$, $T_2, T_3, \ldots$ that decreases toward 0 is selected. A predetermined number $n_i$ of Metropolis updates are performed using $V_i(A) = (kT_i)^{-1}V^*(A)$, $i = 1, 2, \ldots$. Usually, it is necessary to make $\{n_i\}$ divergent. If $\{n_i\}$ diverges sufficiently rapidly (the speed depending on certain properties of $V^*$, of course), then $V^*$ will be frozen at its maximum value with probability 1 in the limit as $n \to \infty$. (See Kirkpatrick [3], and Hajek, [4].)

*References*

[1] C. J. Preston, Gibbs States on Countable Sets, Cambridge Tracts, Cambridge University Press, 1974.

[2] B. Hajek and T. Berger, "A Decomposition Theorem for Binary Markov Random Fields, *Annals of Probability*, vol. 15, no. 3, pp. 1112-1125, July 1987.

[3] S. Kirkpatrick, et al.,Optimization by Simulated Annealing, *Science*, vol. 220, pp. 671-679, 1983.

[4] B. Hajek, Cooling Schedules for Simulated Annealing, preprint, University of Illinois, Coordinated Science Laboratory, submitted to *Mathematics of Operations Research*.

# LINEAR COMPLEXITY IN CODING THEORY

James L. Massey[1] and Thomas Schaub[2]

[1]Institute for Signal and Information Processing
Swiss Federal Institute of Technology
CH-8092 Zurich, Switzerland

[2]Central Laboratory
Landis & Gyr
CH-6301 Zug, Switzerland

## Abstract

The linear complexity of sequences is defined and its main properties reviewed. The linear complexity of periodic sequences is examined in detail and an extensive list of its properties is formulated. The discrete Fourier transform (DFT) of a finite sequence is then connected to the linear complexity of a periodic sequence by Blahut's theorem. Cyclic codes are given a DFT formulation that relates their minimum distance to the least linear complexity among certain periodic sequences. To illustrate the power of this approach, a slight generalization of the Hartmann-Tzeng lower bound on minimum distance is proved, as well as the Bose-Chaudhuri-Hocquenghem lower bound. Other applications of linear complexity in the theory of cyclic codes are indicated.

## I. Introduction

The concept of linear complexity has proved to be very useful in cryptography, particularly in the theory of stream ciphers. The thesis of this paper is that this concept is at least as useful in coding theory, particularly in the theory of cyclic codes.

Section II of this paper gives the definition of linear complexity for an arbitrary sequence, finite or semi-infinite. The special case of periodic semi-infinite sequences is treated in detail in Section III, where an extensive list of linear complexity properties are collected. Certain linear-independence properties of N-tuples, needed in the sequel, are derived in Section IV. A result, called Blahut's Theorem, which relates the linear complexity of a periodic sequence to the discrete Fourier transform (DFT) of its first period, is reviewed in Section V. Section VI shows that the minimum distance of a cyclic code can be underbounded by the least linear complexity of certain periodic sequences. This sets the stage for Section VII in which the results of Sections III and IV are applied to obtain the Bose-Chaudhuri-Hocquenghem (BCH) lower bound and a slightly-strengthened version of the Hartmann-Tzeng lower bound on the minimum distance of a cyclic code. Some other applications of linear complexity in coding theory are given, together with concluding remarks, in Section VIII.

## II. Linear Complexity

We shall write $b^n$ to denote a sequence $(b_0, b_1, \ldots, b_{n-1})$ of length n whose components lie in some specified field F. We shall allow n = 0, in which case $b^0$ is the empty sequence, and we shall allow $n = \infty$, in which case $b^\infty$ is the semi-infinite sequence $(b_0, b_1, b_2, \ldots)$. The linear complexity of $b^n$, here denoted $\Lambda(b^n)$, can then be defined as the smallest nonnegative integer L such that there exists $c_1, c_2, \ldots, c_L$ in F for which

$$b_j + c_1 b_{j-1} + \ldots + c_L b_{j-L} = 0 \text{ for } L \leq j < n, \qquad (1)$$

and as $\infty$ in case no such integer L exists. This definition is "almost" the same as saying that $\Lambda(b^n)$ is the order of the homogeneous linear recursion of least order satisfied by $b^n$ -- the slight difference is that we do not require $c_L$ to be non-zero so that the linear recursion (1) might have order less than L.

It follows immediately from this definition that

$$\Lambda(b^n) = 0 \quad \text{iff} \quad b^n = 0^n, \qquad (2)$$

where $0^n$ here and hereafter denotes the all-zero sequence of length n and where "iff" stands for "if and only if". Another immediate consequence is that, for $0 < n < \infty$,

$$\Lambda(b^n) = n \quad \text{iff} \quad b^{n-1} = 0^{n-1} \quad \text{and} \quad b_{n-1} \neq 0. \qquad (3)$$

The engineering interpretation of $\Lambda(b^n)$ is as the length L of the shortest linear-feedback shift-register (LFSR) that can generate the sequence, see Fig. 1. Such an LFSR is said to be non-singular when $c_L \neq 0$, i.e., when it corresponds to a linear recursion of order L.

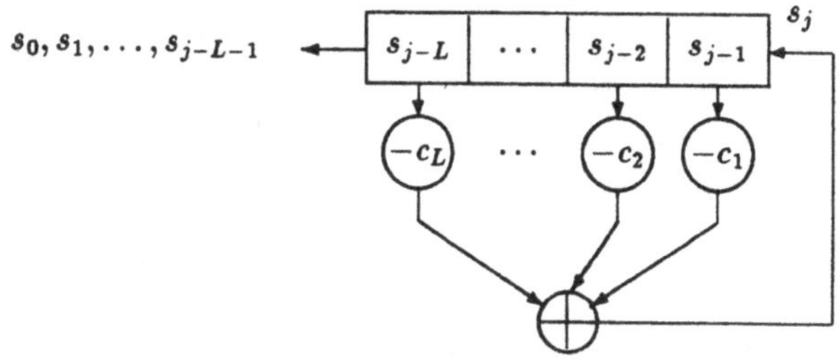


Figure 1: **The linear-feedback shift-register interpretation of the recursion (1)**

Example 1: The sequence $b^5 = (0,1,1,1,1)$ has $\Lambda(b^5) = 2$. The corresponding coefficients in (1) are unique and given by $c_1 = -1$ and $c_2 = 0$; thus, the unique shortest LFSR generating $b^5$ is singular.

There is an efficient algorithm [1] for computing $\Lambda(b^n)$ that also finds the coefficients $c_1, c_2, \ldots, c_L$ in (1) with $L = \Lambda(b^n)$. This algorithm is commonly called the "Berlekamp-Massey" (BM) algorithm. The following fundamental properties of linear complexity are proved in [1]:

Uniqueness Property: The coefficients $c_1, c_2, \ldots, c_L$ in (1) for $L = \Lambda(b^n)$ are unique iff $L \leq n/2$.

Complexity-Change Property: $\Lambda(b^{n+1}) \geq \Lambda(b^n)$ holds for all n, $0 \leq n < \infty$. If $\Lambda(b^{n+1}) \neq \Lambda(b^n)$, then

$$\Lambda(b^{n+1}) = n + 1 - \Lambda(b^n).$$

This complexity-change property was exploited by Rueppel [2] to determine the mean and variance of $\Lambda(b^n)$when $b^n$ is a coin-tossing sequence over $F = GF(2)$, the finite field of two elements, see also [3, pp. 36-41]. Rueppel's book [3] is an excellent source for the uses of linear complexity in cryptography.

For completeness, we state also the following trivial property.

Subsequence Property: If $b^n$ is a subsequence of $a^m$, i.e., if for some integer i $b_j = a_{j+i}$ for $0 \leq j < n$, then $\Lambda(b^n) \leq \Lambda(a^m)$.

## III. Periodic Sequences

Our main interest will be in the linear complexity of periodic sequences. We shall say that the semi-infinite sequence $b^\infty$ is periodic with period N when N is a positive integer such that

$$b_j = b_{j+N} \quad \text{for all } j \geq 0. \tag{4}$$

The smallest N for which a sequence is periodic with period N will be called its fundamental period. Thus $0^\infty$ is periodic with period N for every positive integer N, but its fundamental period is 1. A periodic sequence $b^\infty$ of period N is completely characterized by its first period $b^N = (b_0, b_1, \ldots, b_{N-1})$.

The cyclic left-shift operator S on N-tuples is defined in the manner that $S(b^n) = (b_1, \ldots, b_{N-1}, b_0)$. By extension, $S^2(b^n) = S(S(b^n)) = (b_2, \ldots, b_{N-1}, b_0, b_1)$, etc. The inverse operator is the cyclic right-shift operator $S^{-1}$ with $S^{-1}(b^N) = (b_{N-1}, b_0, b_1, \ldots, b_{N-2})$. The operator S is a linear operator on the vector space of N-tuples over F, and under composition obeys the usual laws of exponents, i.e. $S^iS^j = S^{i+j}$ for any integers i and j. Thus S can be treated as an indeterminate, and polynomials in S with coefficients in F manipulate in the usual manner, e.g., $(S+1)(S-1) = S^2-1$.

If $b^\infty$ is periodic with period N, then (1) is entirely equivalent to

$$b_j + c_1b_{j-1} + \ldots + c_Lb_{j-L} = 0 \quad \text{for} \quad L \leq j < N+L. \tag{5}$$

Equation (5) in turn is equivalent to the single N-tuple equation

$$S^L(b^N) + c_1S^{L-1}(b^N) + \ldots + c_Lb^N = 0^N. \tag{6}$$

Because $S^N$ is the identity operator, the following property follows immediately from (6) and the Uniqueness Property.

Periodic Sequence Property: If $b^\infty$ is periodic with period N then

$$\Lambda(b^\infty) \leq N$$

and there are unique coefficients $c_1, \ldots, c_L$ for which (6) is satisfied with $L = \Lambda(b^\infty)$; moreover,

$$c_L \neq 0.$$

It follows that the linear complexity of a periodic sequence is precisely the order of the homogeneous linear recursion (1) of least order satisfied by the sequence. We write $c(X) = X^L + c_1X^{L-1} + \ldots + c_L$ for the characteristic polynomial of this linear recursion and, when $L = \Lambda(b^\infty)$, we call $c(X)$ the <u>characteristic polynomial of the periodic sequence</u> $b^\infty$. The characteristi polynomial is monic, i.e., its leading coefficient is 1. We see now that the linear complexity of a periodic sequence $b^\infty$ of period N can equivalently be defined as the degree of the smallest degree monic polynomial $c(X)$ such that

$$c(S)(b^N) = 0^N. \tag{7}$$

The following property is fundamental.

<u>Division Property</u>: If $b^\infty$ is periodic with period N and $p(X)$ is any polynomial over F, then

$$p(S)(b^N) = 0^N \quad \text{iff} \quad c(X) \text{ divides } p(X)$$

where $c(X)$ is the characteristic polynomial of $b^\infty$; in particular, $c(X)$ divides $X^N - 1$.

This property is proved by writing $p(X) = q(X)c(X) + r(X)$ where degree $[r(X)] < \text{degree } [c(X)] = \Lambda(b^\infty)$ and noting that $p(S)(b^N) = 0^N$ implies $r(S)(b^N) = 0^N$ and hence $r(X) = 0$. The particular case follows from the fact that $S^N(b^N) = b^N$, i.e. $(S^N-1)(b^N) = 0^N$.

The following property generalizes the well-known "shift-and-add property" of maximum-length sequences [4, p.75].

<u>Generalized Shift-and-Add Property</u>: If $b^\infty$ is periodic with period N, $p(X)$ is any polynomial over F, and $a^\infty$ is the periodic sequence with period N whose first period is $a^N = p(S)(b^N)$, then the characteristic polynomials $c_b(X)$ and $c_a(X)$ of $b^\infty$ and $a^\infty$, respectively, are related by

$$c_b(X) = c_a(X) \gcd[c_b(X), p(X)] \tag{8}$$

(where "gcd" denotes "greatest common divisor"). In particular, $\Lambda(a^\infty) \leq \Lambda(b^\infty)$ with equality iff $\gcd[c_b(X), p(X)] = 1$.

To prove this property, we note that $c(S)(a^N) = c(S)p(S)(b^N)$. From the Division Property, it follows that $c(S)(a^N) = 0^N$ iff $c_b(X)$ divides $c(X)p(X)$. But $c(X) = c_b(X)/\gcd[c_b(X), p(X)]$ is the monic polynomial of smallest degree such that $c(X)p(X)$ is divisible by $c_b(X)$ and hence this polynomial is $c_a(X)$.

The engineering interpretation of the generalized shift-and-add property is that the minimal-length LFSR [with feedback polynomial $c_b(X)$] that generates $b^\infty$ also generates $a^\infty$, any sequence whose first period $a^N$ is a linear combination of $b^N$ and its cyclic shifts.

To the first period $b^N$ of a periodic sequence $b^\infty$ we now associate the <u>circulant matrix</u>

$$M[b^N] = \begin{bmatrix} b_0 & b_1 & \cdots & b_{N-1} \\ b_1 & b_2 & \cdots & b_0 \\ \vdots & \vdots & & \vdots \\ b_{N-1} & b_0 & \cdots & b_{N-2} \end{bmatrix} \tag{9}$$

whose rows are the N-tuples $b^N$, $S(b^N), \ldots, S^{N-1}(b^N)$.

<u>Rank Property</u>: If $b^\infty$ is periodic with period N, then

$$\Lambda(b^\infty) = \text{rank}\,(M[b^N]). \tag{10}$$

This property is proved by noting from (6) that $\Lambda(b^\infty) = L$ is the smallest positive integer such that row L+1 of $M[b^N]$ is linearly dependent on the L previous rows so that the first L rows of $M[b^N]$ are linearly independent, and then noting that (6) also implies

$S^{L+i}(b^N) + c_1 S^{L+i-1}(b^N) + \ldots + c_L S^i(b^N) = 0$ for all $i \geq 0$ so that each subsequent row of $M[b^N]$ is linearly dependent on the L preceeding rows.

By the <u>k-th decimation</u> (where k is a positive integer) of the sequence $b^\infty$, one means the sequence $a^\infty = (b_0, b_k, b_{2k}, \ldots)$. When $b^\infty$ is periodic with period N then so of course is its k-th decimation $a^\infty$, but the latter may have a smaller fundamental period; moreover, $a^N = [b_0, b_k, b_{2k}, \ldots, b_{(n-1)k}]$.

<u>Decimation Lemma</u>: If $a^\infty$ is the k-th decimation of $b^\infty$, a periodic sequence with period N, then

$$\Lambda(a^\infty) \leq \Lambda(b^\infty), \tag{11}$$

with equality if gcd(k,N) = 1.

To prove this property, we note first that $M[a^N]$ is the submatrix of the kN × kN matrix

$$\left[\begin{array}{c|c|c|c} M[b^N] & M[b^N] & \cdots & M[b^N] \\ \hline M[b^N] & M[b^N] & \cdots & M[b^N] \\ \hline \vdots & \vdots & \vdots & \vdots \\ \hline M[b^N] & M[b^N] & \cdots & M[b^N] \end{array}\right]$$

obtained by deleting all rows except rows 1, k+1, 2k+1, ..., (N-1)k+1 and all columns except columns 1, k+1, 2k+1, ..., (N-1)k+1. Thus the rank of $M[a^N]$ cannot exceed the rank of this larger matrix, which of course has the same rank as $M[b^N]$. Thus (11) follows from the Rank Property. When gcd(k,N) = 1, then there is a positive integer i such that ik = QN + 1 for some integer Q. It follows that the i-th decimation of $a^\infty$ is then $(b_0, b_{ik}, b_{2ik}, \ldots)$ $= (b_0, b_1, b_2, \ldots) = b^\infty$ and hence that $\Lambda(b^\infty) \leq \Lambda(a^\infty)$ holds as does also $\Lambda(a^\infty) \leq (b^\infty)$; thus $\Lambda(a^\infty) = \Lambda(b^\infty)$.

## IV. Some Independence Properties

The following independence property was already obtained in our proof of the Rank Property.

Independence Property I: If $b^\infty$ is periodic with period N, then $b^N$, $S(b^N)$, ..., $S^{i-1}(b^N)$ are linearly independent N-tuples iff $i \leq \Lambda(b^\infty)$.

The next independence property is slightly more subtle and is the key to the generalized Hartmann-Tzeng bound of Section VII.

Independence Property II: If $b^\infty$ is periodic with period N and if k is a positive integer such that $\gcd(k,N) < \Lambda(b^\infty)$, then $b^N$ and $S^k(b^N)$ are linearly independent N-tuples.

To prove this property, we suppose that $b^N$ and $S^k(b^N)$ are linearly dependent or, equivalently, that for some $\gamma \neq 0$

$$S^k(b^N) = \gamma b^N.$$

It follows that, for any positive integer i,

$$S^{ik}(b^N) = \gamma^i b^N.$$

Choosing i as a positive integer such that, for some integer j, $\gcd(k,N) = ik + jN$ then gives

$$S^{\gcd(k,N)-jN}(b^N) = S^{\gcd(k,N)}(b^N) = \gamma^i b^N.$$

It follows that $S^{\gcd(k,N)}(b^N)$ and $b^N$ are linearly dependent, which by Independence Property I implies that $\gcd(k,N) \geq \Lambda(b^\infty)$, as was to be shown.

## V. Discrete Fourier Transforms and Blahut's Theorem

We now suppose that the field F contains a primitive N-th root of unity $\alpha$, i.e., an element $\alpha$ such that $\alpha^N = 1$ but $\alpha^i \neq 1$ for $i = 1,2,\ldots,N-1$. The discrete Fourier transform (DFT) of the F-ary sequence $b^N$ is defined to be the F-ary sequence $B^N$ where

$$B_i = \sum_{j=0}^{N-1} b_j \alpha^{ij} \tag{12}$$

for $i = 0,1,\ldots,N-1$, where $\alpha$ is some specified primitive N-th root of unity in F.

From the fact that, for any integer k,

$$\sum_{i=0}^{N-1} (\alpha^k)^i = \begin{cases} N^*, & \text{if N divides k} \\ 0, & \text{otherwise} \end{cases} \tag{13}$$

[which, in the nontrivial case where N does not divide k, is proved by multiplying the sum by $\alpha^k-1$ (which cannot be 0) to obtain $\alpha^{Nk}-1 = 0$] where $N^*$ on the right in (13) means the sum of N 1's in the field F, one finds by direct calculation the <u>inverse DFT</u> relation

$$b_j = \frac{1}{N^*} \sum_{i=0}^{N} B_i \alpha^{-ij} \tag{14}$$

for $j = 0,1,\ldots,N-1$. [To obtain (14), we used the fact that $N^* \neq 0$, which follows from the fact that $X^N-1$ has N distinct roots (viz., $1,\alpha,\ldots,\alpha^{N-1}$) and hence that $X^N-1$ must be relatively prime to its formal derivative $N^*X^{N-1}$.]

If one now uses (12) to define $B_i$ for all nonnegative integer i, one sees that

$$B_{i+N} = \sum_{j=0}^{N-1} b_j \alpha^{(i+N)j} = \sum_{j=0}^{N-1} b_j \alpha^{ij} = B_i$$

and hence that $B^\infty$ is periodic with period N. Similarly, if one uses (13) to define $b_j$ for all nonnegative integers j, one finds that $b^\infty$ is periodic with period N. Thus, it is natural to consider that <u>$b^N$ and $B^N$ are the first periods of the periodic sequences $b^\infty$ and $B^\infty$ with period N defined by (14) and (12) for all $i \geq 0$ and all $j \geq 0$, respectively,</u> as we hereafter so consider.

By the <u>Hamming weight</u> $W(b^N)$ of an N-tuple $b^N$ is meant the number of its components that are different from 0.

The following fundamental result, which is implicit in the work of Blahut [5], provides a link between linear complexity and the DFT.

<u>Blahut's Theorem:</u> If $B^N$ is the DFT of $b^N$, then

$$\Lambda(b^\infty) = W(B^N)$$

and

$$W(b^N) = \Lambda(B^\infty).$$

To prove Blahut's Theorem, we first define $A_j^\infty$ to be the non-zero periodic sequence with first period

$$A_j^N = (1, \alpha^j, \alpha^{2j}, \ldots, \alpha^{(N-1)j}) \tag{15}$$

for each j, $0 \leq j < N$. We note that $\alpha^j A_j^N = S(A_j^N)$ and hence that, for any polynomial p(X),

$$p(S)(A_j^N) = p(\alpha^j)A_j^N. \tag{16}$$

If $b^N = 0^N$, then $B^N = 0^N$ and Blahut's Theorem holds trivially. If $b^N \neq 0^N$, then let $b_{n(j)}$ for $j = 1,2,\ldots,W(b^N)$ be the non-zero components of $b^N$. It follows from (12) and (15) that

$$B^N = \sum_{j=1}^{W(b^N)} b_{n(j)} A_{n(j)}^N$$

and thus from (16) that

$$p(S)(B^N) = \sum_{j=1}^{W(b^N)} b_{n(j)} p(\alpha^{n(j)}) A_{n(j)}^N. \tag{17}$$

But (17) implies that

$$c(X) = \prod_{j=1}^{W(b^N)} (X-\alpha^{n(j)}) \tag{18}$$

is precisely the characteristic polynomial of $B^\infty$, as follows from the facts that (1) $c(\alpha^{n(j)}) = 0$ for $j = 1,2,\ldots,W(b^N)$ so that the sum on the right of (17) vanishes for $p(S) = c(S)$, and (2) $p(S)B^N = 0^N$ is impossible if $p(\alpha^{n(j)}) \neq 0$ for some j, $1 \leq j \leq W(b^N)$, since then if one multiplies $p(S)B^N$ by $S-\alpha^{n(k)}$ for all $k \neq j$ one must, according to (16), obtain a non-zero result. Thus, $\Lambda(B^\infty)$ = degree $c(X) = W(b^N)$. The other half of Blahut's Theorem can be similarly proved.

A polynomial formulation of the DFT will prove useful in the sequel. With the N-tuple $b^N$ we can and will associate in a one-to-one manner the polynomial

$$b(X) = b_0 + b_1X + \ldots + b_{N-1}X^{N-1}.$$

The DFT relation (12) can then be written simply as

$$B_i = b(\alpha^i). \qquad (19)$$

## VI. Cyclic Codes and Linear Complexity

An F-ary linear (N,K) code, where $N \geq 1$ and $1 \leq K \leq N$ is a K-dimensional subspace of the vector space $F^N$ of all F-ary N-tuples $b^N$. It is customary to take F to be a finite field GF(q), but this restriction is not necessary and will not be enforced here. The minimum distance, $d_{min}$, of such a code is the smallest number of positions in which two distinct codewords differ. Because the codewords form a subspace, it follows that

$$d_{min} = W_{min}$$

where $W_{min}$ is the minimum Hamming weight among the non-zero codewords.

An (N,K) cyclic code is a linear (N,K) code such that whenever $b^N$ is a codeword so also is its cyclic shift $S(b^N)$. It is easy to show (cf.[6, pp. 97-98]) that an (N,K) cyclic code is characterized completely by its generator polynomial g(X), which is the unique polynomial over F of degree N-K such that $b^N$ is a codeword iff g(X) divides b(X); moreover, g(X) divides $X^N-1$.

Let $g^N$ be the N-tuple corresponding to g(X), and let $g^\infty$ be the periodic sequence with first period $g^N$. As we shall soon see, the locations of the zeroes in the sequence $g^\infty$ can be used to obtain interesting lower bounds on $d_{min}$.

We will always assume that the field F contains a primitive N-th root of unity $\alpha$ that we shall use to define the DFT $B^N$ of $b^N$. Because $\alpha$ is a primitive N-th root of unity, it follows that

$$X^N-1 = \prod_{i=0}^{N-1} (X-\alpha^i)$$

and hence that g(X) has N-K distinct roots in F. [The reader may be aware that it is more usual to demand only that a primitive N-th root of unity exist in some extension field E of F, but this is no real generalization since g(X) can also be considered as a polynomial with coefficients in E and the resultant E-ary cyclic code generated by g(X) has the same dimension K and same minimum distance $d_{min}$ as the F-ary cyclic code generated by this same g(X).] It follows that exactly N-K components of the DFT $G^N$ of $g^N$ are zeroes.

If $b^N$ is a codeword in the cyclic code generated by g(X) then b(X) = a(X)g(X) for some polynomial a(X) of degree less than K. It follows then from the DFT relation (19) that

$$B_i = A_i G_i$$

for all $i \geq 0$, where A is the periodic sequence whose first period $A^N$ is the DFT of the N-tuple $a^N$ corresponding to a(X). In particular, $G_i = 0$ implies $B_i = 0$. Moreover, because the polynomials a(X) of degree less than K form a K-dimensional vector space, so also do the N-tuples $B^N$ resulting from codewords, and hence this latter space must be exactly the K-dimensional space of all N-tuples $B^N$ with 0's in the N-K locations corresponding to the zeroes of g(X). Moreover, Blahut's theorem specifies that $W(b^N) = \Lambda(B^\infty)$ where $B^\infty$ is the periodic sequence with first period $B^N$. We have thus proved the following result, which is the main theorem of this paper.

Zero-Location Theorem: The minimum distance $d_{min}$ of the (N,K) cyclic code generated by g(X) is equal to the minimum linear complexity $\Lambda(B^\infty)$ among all periodic sequences $B^\infty$ whose first period $B^N$ is a non-zero N-tuple with a 0 in each component where $G^N$, the DFT of $g^N$, contains a 0.

## VII. Two Bounds on $d_{min}$ for Cyclic Codes

Suppose that $G^\infty$ contains d-1 consecutive zeroes or, equivalently from (19), that $g(\alpha^i) = 0$ for $i = m_0, m_0+1, \ldots, m_0+d-2$ for some nonnegative integer $m_0$. Then every $B^\infty$ with first period $B^N$ corresponding to a non-zero codeword $b^N$ is a non-zero periodic sequence containing d-1 consecutive zeroes. It follows from (3) and the Subsequence Property of Section II that $\Lambda(B^N) \geq d$. The Zero-Location Theorem now gives the usual Bose-Chaudhuri-Hocquenghem (BCH) bound on $d_{min}$.

BCH Bound: The (N,K) cyclic code generated by g(X) for which $g(\alpha^i) = 0$ for $i = m_0, m_0+1, \ldots, m_0+d-2$ has minimum distance $d_{min} \geq d$.

As a less trivial illustration of the power of the linear complexity approach to bounding $d_{min}$ for a cyclic code, we now prove the Hartmann-Tzeng bound [7] in its slightly strengthened version as given in [8].

Strengthened Hartmann-Tzeng Bound: The (N,K) cyclic code generated by g(X) for which $G^\infty$ contains a subsequence

$$0^m\Delta^{n-m}0^m\Delta^{n-m}\ldots 0^m\Delta^{n-m} \qquad (20)$$

of length ns, where $1 \leq m \leq n$, where $s \geq 1$, where each $\Delta^{n-m}$ symbolizes a sequence of n-m arbitrary elements of F, and where $\gcd(n,N) \le m$, has minimum distance $d_{min} \ge m+s$.

For convenience, we shall denote the sequence (20) by $[0^m\Delta^{n-m}]^s$. If $b^N$ is a non-zero codeword in the cyclic code generated by g(X), then $B^\infty$ is a non-zero periodic sequence of period N that also contains the subsequence $[0^m\Delta^{n-m}]^s$. By the BCH bound argument, $\Lambda(B^\infty) \ge m+1$. It follows then from Independence Property II that $B^N$ and $S^n(B^N)$ are linearly independent, because $\gcd(n,N) \le m$. Thus, $A^N = B^N - \gamma S^n(B^N)$ for every $\gamma \varepsilon F$ is nonzero and $A^\infty$ contains the subsequence $[0^m\Delta^{n-m}]^{s-1}$. We will show that, for some choice of $\gamma$ in F, $\Lambda(A^\infty) \le \Lambda(B^\infty)-1$. Iteration of this argument will then show that $\Lambda(C^\infty) \le \Lambda(B^\infty) - (s-1)$ where $C^\infty$ is a non-zero periodic sequence containing the subsequence $0^m\Delta^{n-m}$. By the BCH argument, $\Lambda(C^\infty) \ge m+1$ which will complete the proof.

It remains only to show that $\gamma$ can be chosen so that $A^N = B^N - \gamma S^n(B^N)$ gives $\Lambda(A^\infty) \le \Lambda(B^\infty)-1$. We first recall from the Division Property of Section II that the characteristic polynomial $c_B(X)$ of $B^\infty$ divides $X^N-1$, and hence there is some i such that $\alpha^i$ is a root of $c_B(X)$. But $\alpha^i$ is also a root of $p(X) = 1-\gamma X^n$ for the choice $\gamma = \alpha^{-in}$. Thus, for this choice of $\gamma$, $\gcd[C_B(X),p(X)]$ has degree at least 1. But $A^N = p(S)B^N$ so the Generalized Shift-and-Add Property of Section II now shows that the characteristic polynomial of $A^\infty$ has degree at least one less than that of $B^\infty$ or, equivalently, that $\Lambda(A^\infty) \le \Lambda(B^\infty)-1$.

## VIII. Concluding Remarks

It follows from the Zero-Location Theorem of Section VI that every lower bound on the minimum distance of a cyclic code can be viewed as a lower bound on the linear complexity of a non-zero periodic sequence whose first period has 0's at the locations corresponding to the roots of the generator polynomial of the code, and conversely. Indeed, we have been able to formulate simple linear complexity arguments similar to those of Section VII to obtain among other results all the lower bounds on $d_{min}$ given in the recent comprehensive paper of van Lint and Wilson [8]; the details will be given in a later paper.

We have focused our attention on periodic sequences in this paper, but we should mention that there is at least one application in coding theory of the linear complexity of finite sequences, namely the following theorem due to Matt and Massey [9]. This theorem is stated in terms of the parity-check polynomial $h(X) = (X^n-1)/g(X)$ of the cyclic code and uses the notation $\lfloor . \rfloor$ to denote the largest integer not greater than the enclosed number.

Theorem: The maximum integer b such that the (N,K) cyclic code with $d_{min} \geq 3$ and parity-check polynomial $h(X) = X^K + h_1 X^{K-1} + \ldots + h_{K-1} X + h_K$ can correct all bursts of length b or less is

$$\min(L_1, L_2, \ldots, L_{K-\delta})$$

where $\delta$ is 0 or 1 according as N-K is even or odd, respectively, and where $L_i$ is the linear complexity of the subsequence of length N-K beginning at position i of the length N sequence

$$0^{m-1}, 1, h_1, \ldots, h_{K-1}, h_K, 0^{N-k-m}$$

and $m = \lfloor (N-K)/2 \rfloor$.

This theorem is of some practical interest as, coupled with the Berlekamp-Massey algorithm [1] to compute the indicated linear complexity, it provides the presently-most-efficient computational method to calculate the burst-correcting limit of a cyclic code, cf. [10]. Its proof, however, seems to offer no particular insight into the underlying reason for this connection between burst-correcting-limit and linear complexity. It seems likely that a stronger and more transparent result could be proved.

## References


[1] J.L. Massey, "Shift-Register Synthesis and BCH Decoding", IEEE Trans. Info.Th., vol. IT-15, pp. 122-127, Jan. 1969.

[2] R. Rueppel, "Linear Complexity and Random Sequences", in Advances in Cryptology - EUROCRYPT 85, (Ed. F. Pichler), Lect. Notes in Comp. Sci. No. 219. Heidelberg and New York: Springer, 1986, pp. 167-188.

[3] R. Rueppel, Analysis and Design of Stream Ciphers. Heidelberg and New York: Springer, 1986.

[4] S.W. Golomb, Shift Register Sequences. San Francisco: Holden-Day, 1967.

[5] R.E. Blahut, "Transform Techniques for Error-Control Codes", IBM J. Res. Devel., vol. 23, pp. 299-315, 1979.

[6] R.E. Blahut, Theory and Practice of Error Control Codes. Reading, MA: Addison-Wesley, 1983.

[7] C.R.P. Hartmann and K.K. Tzeng, "Generalizations of the BCH Bound", Info. and Control, vol. 20, pp. 489-498, 1972.

[8] J.H. van Lint and R.M. Wilson, "On the Minimum Distance of Cyclic Codes", IEEE Trans. Info. Th., vol. IT-32, pp. 23-40, Jan. 1986.

[9] H.J. Matt and J.L. Massey, "Determining the Burst-Correcting Limit of Cyclic Codes", IEEE Trans. Info. Th., vol. IT-26, pp. 289-297, May 1980.

[10] T. Kasami, "Comments on 'Determining the Burst-Correcting Limit of Cyclic Codes'", IEEE Trans. Info. Th., vol. IT-27, p. 812, Nov. 1981.

# K-PERMUTIVITY

Gilbert ROUX
UER de mathématiques, Université de PARIS 6
75005 PARIS

**Résumé** On dit qu'un tableau T de n colonnes est k-permutif si et seulement si pour tout k-uplet de colonnes $\{i_1,..,i_k\}$, le sous tableau extrait de T, constitué par ces colonnes contient en lignes toutes les permutations de $\{i_1,..i_k\}$ . Pour n et k fixés, on va encadrer le nombre p(n,k) de lignes d'un tel tableau. On obtient:

$A_n^{k-1} \leqslant p(n,k) \leqslant A_n^{k-1}.d_{k+1} / k!$ où $A_n^k$ est le nombre d'injections d'un ensemble de cardinal p dans un ensemble de cardinal n, et $d_{k+1}$ le nombre de permutations d'un ensemble de k+1 élèments, sans point fixe.

**Summary** An n-column array T is k-permutive if, for each k-tuple of columns $\{i_1,..i_k\}$,the sub-array constituted by these columns contains as rows all the k! permutations of $\{i_1,..,i_k\}$.When n and k are fixed, we will give lower and upper bounds on the number p(n,k) of rows of such an array. We obtain:

$A_n^{k-1} \leqslant p(n,k) \leqslant A_n^{k-1}.d_{k+1} / k!$ where $A_n^k$ is the number of injections of a set of cardinality k to a set of cardinality n, and $d_{k+1}$ is the number of derangments of $\{i_1,...,i_k\}$.

## Introduction

Recently , many problems consisting in minimizing the number of rows of an array with n columns appeared. For instance k-surjectivity [C.K.M.Z] : minimize the number of rows of a n-columns array T such that for each k-tuple $\{i_1,..,i_k\}$ of $\{1,..,n\}$ $k \leqslant n$ , the sub-array constituted with columns of T of indice $i_1,...,i_k$ contains as rows all the binary k-vectors.

Another example: k-conflicts without feedback [ K.G] ; here, the same array must contain all the binary n-vectors with weight 1.

Here, we shall deal with in another class of such problems: the case when the elements of the array T are not elements of a fixed set but depend on n. The example treated is k-permutivity [P.J.S]: minmize the number of rows of an n column array T constructed on $\{1,2,..,n\}$ such that, for all k-tuple of columns, the sub-array constituted by these columns contains as rows all the k! permutations of the indices of these columns.

Let T be an array of r rows and n columns. T is called <u>k-permutive</u> ( $1 \leq k \leq n$ ) if and only if, for each k-tuple of columns with indices $i_1,..,i_k$, the rxk array constituted by these columns contains all the k! permutations of $\{i_1,...,i_k\}$

for instance:

$$T = \begin{matrix} 1 & 2 & 3 & 4 & 5 \\ 5 & 4 & 3 & 2 & 1 \\ 4 & 3 & 2 & 1 & 5 \\ 3 & 2 & 1 & 5 & 4 \\ 2 & 1 & 5 & 4 & 3 \\ 1 & 5 & 4 & 3 & 2 \end{matrix} \quad \text{is a 2-permutive array}$$

We can also speak of k-permutivity in terms of permutations: let us call k-permutation of a set E, each permutation of a part of E with cardinality k. So, a set $\mathcal{C}$ of permutations of $\{1,...,n\}$ is called k-permutive if and only if, for each k-permutation s defined on a part $\{x_1,...,x_k\}$ of $\{1,..,n\}$, then s is the restriction of a permutation of $\mathcal{C}$ to $\{x_1,...,x_k\}$, this means, there exists an element t of $\mathcal{C}$ such that: for each i element of $\{1,..,k\}$ , $t(x_i) = s(x_i)$.

Let us note <u>p(n,k) the minimum number</u> of rows of a k-permutive array with n columns.

The problem is to find values of p(n,k) for all n and k. It first appeared in a paper of Peter J. SLATER [P.J. S] in 1979, under the title:" how few n-permutations contain all k-permutations?".

SLATER has shown that $p(2m-1,2) = p(2m,2) = 2m$ for all m, and that $p(n,n-1) = n!-d_n$ where $d_n$ is the number of derangments of $\{1,2,..,n\}$ ( a derangment of $\{1,2,..,n\}$ is a permutation of $\{1,2,..,n\}$ without fixed point ).

In the present paper, upper and lower bounds of p(n,k) will be found, and it will be shown that $p(n,k) = \Theta(n^{k-1})$ ( $f(n) = \Theta(g(n))$ means that for n sufficiently large, there exist two constants c and c' such that $c|g(n)| < |f(n)| < c'|g(n)|$)

<u>Notation</u>

* $A_n^p$ is the number $n(n-1)...(n-p+1)$ .
* $\binom{n}{p}$ the number of p subsets of a set of cardinal n.
* $\lfloor x \rfloor$ is the lower integer part of x, i.e. the integer n such that $n \leq x < n+1$

  $\lceil x \rceil$ is the upper integer part of x, i.e. the integer n such that $n-1 < x \leq n$

First, it is easy to see that $\boxed{p(n,k) \leq A_n^k}$

Indeed there are $A_n^k$ k-permutations of {1,2,..,n}. So, if s is one of these k-permutations, defined on $\{x_1,...,x_k\}$, let t be the permutation of {1,2,..,n} such that: $\begin{cases} x \in \{x_1,..,x_k\}, & t(x)=s(x) \\ x \notin \{x_1,..,x_k\}, & t(x)=x \end{cases}$.

Then the set $\mathcal{C}$ of such permutations t is clearly k-permutive.

1) lower bound.

We will show that, for all n and k, $p(n,k) \geqslant A_n^{k-1}$ .

* the case k=3

Let T be a 3-permutive array, and suppose that rows of T are such that:

$$T = \begin{bmatrix} 1 & \\ \vdots & \\ 1 & \\ 2 & \\ \vdots & \\ 2 & \\ \vdots & \\ n & \\ \vdots & \\ n & \end{bmatrix}$$

Let $T_i$ be the array constituted by the rows whose first element is i .

Let $T_1^*$ the array constituted by the columns of indices 2,3,..,n of $T_1$

If we substract 1 to each element of $T_1^*$, $T_1^*$ is a 2-permutive array of n-1 columns. Hence $T_1^*$ and $T_1$ have at least n-1 rows .

Now, consider $T_i$ with $i \neq 1$. Let s be a permutation defined on {1,i,j} $j \neq 1$ and $j \neq i$, such that s(1)=i. There exists a permutation t, element of $T_i$, which covers s ( we say that a permutation t covers a permutation s defined on $\{i_1,..,i_k\}$ if and only if, for each element x element of $\{i_1,...,i_k\}$ , t(x)=s(x).

So $T_i$ contains a permutation t such that t(i)=1 ( When s=(i,1,j))

$T_i$ contains a permutation t such that t(i)=j ( when s=(i,j,1))

As j can take all values different from 1 and i, $T_i$ has at least n-1 rows

$$p(n,3) \geqslant n(n-1) .$$

* general case by induction on k.

Hypothesis ( $H_k$ )

1) if $\mathcal{C}_{i,u}$ is a set of permutations such that every k-permutation s of {1,2,..,n} which verifies s(u)=i is covered by an element of $\mathcal{C}_{i,u}$, then $\mathcal{C}_{i,u}$ has at least $A_{n-1}^{k-2}$ rows.

2) $p(n,k) \geqslant A_n^{k-1}$ .

* These hypothesis are verified when k=3 .

* Suppose now $H_3, H_4, .., H_{k-1}$ true .

To prove the first part of $H_k$ , for limpidity of the demonstration,we will set u=1, and so we set $T_{i,1} = T_i$ .

So let $T_i$ be an array such that every k-permutation of {1,..,n} which verifies s(1)=i is a restriction of an element of $T_i$ .

* if i=1 , each (k-1)-permutation of {2,3,..,n} is covered by an element of the sub-array $T_i^*$ of $T_i$ constituted by the columns of indices 2,3,..,n.
Hence $T_i^*$ is (k-1)-permutive. So by 2), $T_1$ has at least $A_{n-1}^{k-2}$ rows.

* if $i \neq 1$ , let $L_j$ be the array constituted by the rows of $T_i$ whose element of the $j^{th}$ column equals 1 , and $a_j$ the number of rows of $L_j$ ( for $j\neq 1$) .

So, if $L_i^*$ is the array obtained after eliminating the columns of $L_i$ of indice 1 and i , it is clear that : $L_i^*$ is (k-2)-permutive with n-2 columns, and so, $a_i \geqslant p(n-2,k-2) \geqslant A_{n-2}^{k-3}$ .

Consider now $a_j$ when $j\neq 1$ and $j\neq i$. We can suppose that j=2 for limpidity of the demonstration.

Let $L'_2$ be the array obtained by eliminating column of indice 1 and by substituting the elements 1 of column of indice 2 by i. If s' is a (k-1)-permutation of {2,3,.,n} such that s'(2)=i then s' is covered by an element of $L'_2$ . Indeed:
if s is the k-permutation of {1,2,..,n} such that s(1)=i , s(2)=1
and for $x\neq 1$ , $x\neq 2$ s(x)= s'(x)
s si covered by an element of $T_i$ and hence by an element of $L_2$. If t' is the permutation associated to t and element of $L'_2$ then t' covers s'.
Hence $L'_2$ verifies the hypothesis 1 of $H_{k-1}$( with u=2)

$a_2 \geqslant A_{n-2}^{k-3}$ . $T_i$ has at least $(n-1).A_{n-2}^{k-3} = A_{n-1}^{k-2}$ rows.
First part of $H_k$ is so verified.

Now, if T is a k-permutive array, we can suppose that rows are such that their first elements are written in increasing order. T is the juxtaposition of n arrays that verifie the first hypothesis of $H_k$ .

=> T has at least $n.A_{n-1}^{k-2} = A_n^{k-1}$ rows; part 2 of $H_k$ is verified

<u>Corollary</u> for all n and $k\geqslant 3$ $A_n^{k-1} \leqslant p(n,k) \leqslant A_n^k$

3) <u>Upper bounds</u>

We now need the following result, due to BARANYAI [BA] in graph theory:

> If X is a set of cardinal n, and $K_n^k(X)$ the collection of all the parts of X of cardinal k, then there exists a partition of $K_n^k(X)$ in $a_{n,k} = \left\lceil \binom{n}{k} \Big/ \left\lfloor \frac{n}{k} \right\rfloor \right\rceil$ classes such that two distinct

elements of one class are disjoints sets, and the number $a_{n,k}$ is minimal among such partitions.

Now let $X=\{1,2,..,n\}$ and $\mathcal{E}_1,..,\mathcal{E}_{a_{n,k}}$ be the elements of this partition Denote by $E_{i,1},E_{i,2},...$ the elements of the class $\mathcal{E}_i$.
Let us associate to the class $\mathcal{E}_i$ the array $T_i$ of k! rows and n columns defined as follows:

if $E_{i,j} = \{x_1,..,x_k\}$, then the rows of the sub-array of $T_i$ constituted by the columns of indices $x_1,..,x_k$ are the k! permutations of $x_1,..,x_k$, the first row is identity permutation of $x_1,...,x_k$.

If $j \neq j'$ then $E_{i,j} \cap E_{i,j'} = \emptyset$ , hence we can repeat such a construction for all $(E_{i,j})_{j=1,2,..}$ . Moreover, if $x \notin \bigcup_j E_{i,j}$ , then each element of the column of indice x of $T_i$ is x .

If $i \neq 1$, let $T_i^*$ be the array constituted by the (k!-1) last rows of $T_i$ .

Set now $T = T_1 \cup ( \bigcup_{i=2}^{a_{n,k}} T_i^* )$ . T is k-permutive, by construction.

Hence, $p(n,k) \leqslant (k!-1).a_{n,k} + 1$

<u>Corollary</u>

| if k divides n, $A_n^{k-1} \leqslant p(n,k) \leqslant A_{n-1}^{k-1}.\left( k-\frac{1}{(k-1)!} \right) + 1$ |
|---|

Hence

| $p(n,k) = \Theta( n^{k-1} )$ |
|---|

We will now inprove the constuction when k<n-2.

* Let $A_i$ be the set of permutations of $\{1,2,..,n\}$ with exactly n-i fixed points. $Card(A_i) = \binom{n}{i}.d_i$ .

* Let T be the array with $d_k$ rows and n columns obtained by a construction similar to the one described to find the upper bound, but writing as rows the $d_k$ k-derangments of $\{x_1,x_2,..,x_k\}$ .

Set now $T = ( \bigcup_{i=2}^{k-1} A_i ) \cup ( \bigcup_{i=1}^{a_{n,k}} T_i )$ .Then T is permutive. Indeed:

* if s is a k-permutation with k-i fixed points, $i \neq 0$ and $i \neq k$ , then s is covered by an element of $A_i$.
* if s is an identity k-permutation, then s is covered by an element of $A_2$ , because $k \leqslant n-2$ .
* if s is a k-permutation without fixed point, defined on $\{x_1,.x_k\}$ then there exists a class $\mathcal{E}_i$ (Baranyai) and an element $E_{i,j}$ of this class such that $E_{i,j}=\{x_1,..,x_k\}$. s is covered by an element of $T_i$ .

Hence:

$$p(n,k) \leqslant d_k . \left\lceil \binom{n}{k} \Big/ \left\lfloor {n \atop k} \right\rfloor \right\rceil + \sum_{i=2}^{k-1} \binom{n}{i} . d_i \qquad k\leqslant n-2$$

If k divides n, then

$$p(n,k) < \binom{n}{2} . d_2 + \binom{n}{3} . d_3 + \ldots + \binom{n}{k-1} . d_{k-1} + \binom{n-1}{k-1} . d_k = X ,$$

$$X = \frac{A_n^{k-1}}{(k-1)!} . \left( d_k . \frac{n-k+1}{n} + d_{k-1} + d_{k-2} . \frac{k-1}{n-k+2} + d_{k-3} . \frac{(k-1)(k-2)}{(n-k+2)(n-k+3)} + .. \right.$$
$$\left. \ldots d_2 . \frac{(k-1)(k-2)\ldots 3}{(n-k+2)\ldots(n-2)} \right)$$

$$= \frac{A_n^{k-1}}{(k-1)!} . \left( d_k + d_{k-1} + \underbrace{\left( \frac{1-k}{n} . d_k + \frac{k-1}{n-k+2} . d_{k-2} + \ldots + \frac{(k-1)(k-2)\ldots 3}{(n-k+2)..(n-2)} . d_2 \right)}_{\beta_{n,k}} \right)$$

It can be shown that $\beta_{n,k} \leqslant 0$ [R]. So, $p(n,k) \leqslant \frac{A_n^{k-1}}{(k-1)!} . ( d_k + d_{k-1} )$

which gives,

| |
|---|
| for $k\geqslant 3$ , $k\leqslant n-2$ , k divides n |
| $A_n^{k-1} \leqslant p(n,k) \leqslant A_n^{k-1} . (d_{k+1} / k!)$ |

[R.]

<u>conclusion</u> We can inprove the lower bound of p(n,k) by resolving linear programming problems . For instance, in the case k=3 we obtain:

$$2\binom{n}{3} \Big/ \left| {n \atop 3} \right| + 3\binom{n}{3} \Big/ i(n-2i) \leqslant p(n,3) \leqslant 2 \left\lceil \binom{n}{3} \Big/ \left\lfloor {n \atop 3} \right\rfloor \right\rceil + \binom{n}{2} \qquad \text{[R.]}$$

with $i = \lfloor (n+2)/4 \rfloor$ .

---

<u>References</u>


[BA] BARANYAI "On the factorisation of the complete uniform hypergraph" Bolyai J.Mat. Tarsulat, Budapest & North Holland, 1975 p:91-108.

[C.K.M.Z] CHANDRA-KOU-MARKOWSKY-ZAKS "On sets of Boolean n-vectors with all k-projections surjective" IBM Research Report 7/17/81.

[K.G] KOMLOS, GREENBERG "A fast non-adaptative algorithm for conflict resolution in multiple access chanels", Mathematics and statistics Research Center, AT & T Bell Laboratories,NJ 07974 (84)

[P.J.S] SLATER "How few n-permutations contain all possible k-permutations?", Journal of Combinatorial theory, Series A 26, 201 , 1979.

[R] ROUX " k-propriétés dans des tableaux de n colonnes" Thèse d'Université , PARIS 6, ( mars 1987 ).

# ON THE DENSITY OF BEST COVERINGS IN HAMMING SPACES

M. Beveraggi and G. Cohen

## NOTATIONS

We write F for the field with two elements. The Hamming distance between the binary n-tuples (elements of $F^n$) is the number of coordinates where they differ. We write it d(.,.), and set $d(\underline{0},Z) = |Z|$, where $\underline{0}$ is the all-zero vector.
Let C be a binary code of length n, linear or not. We denote by t or t(C) the covering radius of C. That is, t is the smallest integer such that the Hamming spheres of radius t centered at the codewords of C cover $F^n$. The minimum possible cardinality of a code with length n and covering radius 1 will be called K(n,1), and

$$\mu_n = 2^{-n}(n+1)K(n,1)$$

will be the minimal density of a code with covering radius 1 (or covering).

## A FEW PROPERTIES

. It is trivial that $\mu_n \geqslant 1$ holds for all n.

. For $n=2^m-1$, m integer, we get $\mu_n=1$ (realized e.g. by Hamming codes). Hence we get

$$\lim_{n\to\infty} \inf \{\mu_n\} = 1.$$

We shall now deal with lim sup $\{\mu_n\}$. To that end we shall consider the two following constructions.

### Construction 1

Let C be a covering in $F^n$. Then to get from it a covering of $F^{n+1}$, it suffices to append to all codewords in C a "0" or a "1" in both possible ways. This a trivial case of cartesian product CxC' (see [1]), where C' is a [1,1] linear code. Therefore we have $K(n+1,1) \leqslant 2\, K(n,1)$

and $\mu_{n+1} \leqslant \mu_n\, (n+2)(n+1)^{-1}$. (1)

### Remark

Iterating construction 1 from $n=2^m-1$ to $n=2^{m+1}-2$ allows to prove : $\mu_n \leqslant 2$.

We now consider a notion introduced in [3], where it was used to construct a perfect code of length nn'+n+n' by combining two perfect codes of length n and n' respectively.

Definition 1. We call generalized parity function from $F^{nn'}$ to $F^{n+n'}$ the function which associates to vector x the following couple of vectors (P(x),P'(x)), where $P(x)=(p_1(x),p_2(x),\ldots,p_n(x))$, $P'(x)=(p'_1(x),p'_2(x),\ldots,p'_{n'}(x))$

and $p_i(x) = \sum_{j=1}^{n'} x_{ij}$, for $i\in\{1,2,\ldots n\}$

$p'_j(x) = \sum_{i=1}^{n} x_{ij}$, for $j\in\{1,2,\ldots,n'\}$.

Here the components of x in $F^{nn'}$ are ordered as follows :

$$x=(x_{11},\ldots,x_{1n'},x_{21},\ldots x_{n'n}).$$

Example

n=3, n'=2, x=(101001).

Then P(x)=(100), P'(x)=(01), which can be visualized as

$$x = \begin{pmatrix}1 & 0 & 1\\0 & 0 & 1\end{pmatrix}\begin{pmatrix}0\\1\end{pmatrix} = P'(x).$$

P(x)= (1 0 0)

For n'=1, we get the usual parity function.

We now introduce the second construction that we need.

Construction 2

Let C and C' be two codes of length n and n' respectively. Define the code C$\star$C' of length nn'+ n+n' by

$$C\star C' := \{x, c+P(x), c'+P'(x)\},$$

where x,c and c' range over $F^{nn'}$, C and C' respectively.

Proposition 1

If C and C' are coverings, so is $C \star C'$.

Proof

Let $y=(u,v,v')$ be any vector in $F^{nn'+n+n'}$. We distinguish two cases

1) $v+P(u) \in C$ or $v'+P'(u) \in C'$.
   Say $v+P(u)=c \in C$. Let c' be the word in C' which is closest to $v'+P'(u)$. Then $s=(u,c+P(u),c'+P'(u))$ is a word in $C \star C'$ such that

   $d(y,s)=d(v',c'+P'(u))=d(v'+P'(u),c') \leqslant t(C')=1$.

2) $v+P(u) \notin C$ and $v'+P'(u) \notin C'$.
   Then $\exists\ e_i \in F^n$, $e'_j \in F^{n'}$, $e_i$ and $e'_j$ of weight 1,

   such that

   $v+P(u)+e_i=c \in C$

   and

   $v'+P'(u)+e'_j=c' \in C'$.

Let $\tilde{u}$ be the element in $F^{nn'}$ (considered as a tableau nxn') which coincides with u except on position (i,j), where

   $\tilde{u}(i,j) = \overline{u(i,j)}$.

Then $v+P(\tilde{u})=c$, $v'+P'(\tilde{u})=c'$ and $(\tilde{u},c+P(\tilde{u}),c'+P'(\tilde{u}))$ is a word in $C \star C'$ at distance $d(u,\tilde{u})=1$ from y.

More generally, one has

Proposition 2

   $t(C \star C') \leqslant \mathrm{Max}\ \{t(C),t(C')\}$.

Proof

In case 1), it is clear.

For 2) $\exists\ z \in F^n$, $|z| \leqslant t(C)$, $\exists z' \in F^{n'}$, $|z'| \leqslant t(C')$

such that

$v+P(u)+z=c$

$v'+P'(u)+z'=c'$.

Suppose $|z|\geq|z'|$ and denote

Supp $(z) = \{i_1,i_2,\ldots,i_{|z|}\}$ the set of non-zero components of z, and similarly

Supp $(z')= \{j_1,j_2,\ldots j_{|z'|}\}$.

Let $\tilde{u}$ be the tableau coïnciding with u, except for

$\tilde{u}(i_1,j_1)=\overline{u(i_1,j_1)},\ldots \quad ,\tilde{u}(i_{|z'|},j_{|z'|})= \overline{u(i_{|z'|},j_{|z'|})}$.

Then $(\tilde{u},c+P(\tilde{u}),c'+P'(\tilde{u}))=s\in C{*}C'$
and $d(s,y)=d(u,\tilde{u})+d(v,c+P(\tilde{u}))+d(v',c'+P'(\tilde{u}))$

$= |z'|+|z|-|z'|+0=|z|$.

Corollary 1 [3]

If C and C' are perfect one-error-correcting codes, then so is C*C'.

Proof. Recall that a perfect code is a covering and a packing (i.e. a partition of $F^n$). Then this is a straightforward consequence of

$|C{*}C'|= 2^{nn'}|C|.|C'|$

and $t(C{*}C')=1$ (by Proposition 2).

Corollary 2

$$\mu_{nn'+n+n'}\leq\mu_n\cdot\mu_{n'} \tag{2}$$

In particular, setting n'=1, we get

$$\mu_{2n+1}\leq\mu_n. \tag{3}$$

Inequality (3) was already obtained in [2],using another construction. Combining (1) and (3), we can upperbound $\mu_n$ :

Proposition 3

For all n, the minimal density of a covering satisfies

$$\mu_n \leqslant 3/2.$$

Proof Using both (1) and (3), we have proved that for all n :

$$\mu_{2n+1} \leqslant \mu_n \text{ and } \mu_{2n+2} \leqslant \frac{2n+2}{2n+1} \mu_n$$

Then let us consider the intervals of the form : $[2^n(p+1)-1, 2^n(p+2)-2]$. There are two "extremal" ways for reducing $2^n(p+2)-2$ to p : first reduce $2^n(p+2)-2$ to $2^n(p+1)-1$ using construction 1 and reduce $2^n(p+1)-1$ to p using construction 2 (with n'=1) ; secondly using alternatively construction 1 and 2 (i.e. using construction 1 when construction 2 is impossible with n'=1).

With these two methods, we obtain the same inequality :

$$\mu_{2^n(p+2)-2} \leqslant \frac{2^n(p+2)-1}{2^n(p+1)} \mu_p.$$

It is easy to check, but rather tedious, that any way of reducing $2^n(p+2)-2$ combining these two methods gives the same result. Hence we get

$$\mu_N \leqslant \frac{2^n(p+2)-1}{2^n(p+1)} \mu_p \quad \text{for all } N \in [2^n(p+1)-1, 2^n(p+2)-2].$$

So we have :

$$\mu_N \leqslant \frac{p+2}{p+1} \mu_p \quad \text{for all } N \in \bigcup_{n \in N^\star} [2^n(p+1)-1, 2^n(p+2)-2].$$

Then, considering that $\mu_i \leqslant 3/2$ for $i \leqslant 5$; $\mu_6 = \frac{21}{16}$ ; $\mu_7 = 1$ ; $\mu_8 = 9/8$ ; $\mu_9 \leqslant 5/4$ ; $\mu_{10} \leqslant \frac{11}{8}$ ; $\mu_{11} \leqslant 9/8$

and $\mu_{12} \leqslant \frac{39}{32}$ (see [4]), and the following :

$\{1,2,3,4,5\} \bigcup_{6 \leqslant p \leqslant 12} \left( \bigcup_{n \in N} [2^n(p+1)-1, 2^n(p+2)-2] \right) = N^\star$, we obtain the proposition. □

Proposition 3 has the following immediate consequence :

$$\limsup \mu_n \leqslant 3/2.$$

## CONCLUSION

There is no way, using the constructions presented here, to improve the 3/2 bound ; but it seems reasonable to conjecture that lim sup $\mu_n = 1$.

## REFERENCES

[1] G.D. COHEN, M.G. KARPOVSKY, H.F. MATTSON, Jr. and J.R. SCHATZ: "Covering radius-survey and recent results", IEEE Trans. Inform. Theory, vol. IT-31, pp. 328-343, 1985.

[2] G.L. KATSMAN and S.N. LITSYN : Private Communication.

[3] M. MOLLARD : "A generalized parity function and its use in the construction of perfect codes", SIAM J.Alg. Disc. Math., vol 7. n° 1, 113-115, 1986.

[4] A. LOBSTEIN, G.D. COHEN and N.J.A. SLOANE : "Recouvrements d'espaces de Hamming Binaires", C.R. Acad. Sc. Paris, t. 301, Série I, n° 5, pp. 135-138, 1985.

RESUME

Nous présentons des constructions pour des recouvrements d'espaces de Hamming binaires de dimension n par des sphères de rayon 1. Nous montrons que la densité minimale $\mu_n$ de tels recouvrements est au plus 3/2. Le comportement asymptotique de $\mu_n$ quand n tend vers l'infini n'est pas connu.

ABSTRACT

We present constructions for coverings of binary Hamming spaces of dimension n by spheres of radius one. We prove that the minimal density $\mu_n$ of such coverings is at most 3/2. The asymptotical behavior of $\mu_n$ is still unknown.

# THE LEE ASSOCIATION SCHEME

Patrick SOLE

INRIA
Domaine de Voluceau
ROCQUENCOURT - B.P. 105
78153 LE CHESNAY CEDEX
FRANCE

Abstract :

In this paper we undertake a study of the Lee scheme. We give in this context a new proof of Bassalygo's generalization of Lloyd Theorem, and an asymptotic estimate of the number of zeroes of the Lloyd polynomial.

We obtain a recursion on the Lee composition distribution of the translates of a code and deduce from that an upper bound on the covering radius of a code.

We give an algebraic characterization of T-designs in this scheme, which shows that they form a special class of orthogonal arrays.

Introduction :

The Lee metric was first introduced by Lee in [12], wich concerns mainly the topic of perfect codes, as several papers from Golomb [10], [11]. A class of negacyclic perfect codes and an engineering motivation can be found in the classical book by Berlekamp [5].

In his thesis [8], Delsarte shown that the concept of an association scheme was a natural framework to coding theory. He introduced the Hamming scheme to study the Hamming metric, and gave the definition of the Lee scheme.

In this paper we make a comparison between the two schemes, and generalize some properties of the former to the latter.

1. The Lee Metric :

Let $Z_q$ (residues modulo q) be taken as an alphabet. We define the weight of a symbol k by :

$$W_L(k) = \text{Max}\ \{k,\ q-k\}$$

By suitably numbering the vertices of the q-gon we see that $W_L(k)$ is the length of the shortest path from 0 to k on this graph.

Then we equip the cartesian product $Z_q^n$ with a metric by the formula :

$$d_L(x,y) = \sum_{i=1}^{n} W_L\ (|y_i - x_i|).$$

2. The Notion of association scheme :

An association scheme with t classes, in the Bose Mesner sense, [7] on the finite set X, consists of a partition $\underline{R} = (R_0, R_1, \ldots, R_t)$ of X × X satisfying the following axioms :

$A_1$ $R_0 = \{(x,x)\ |x \in X\}$

$A_2$ $R_i^{-1} = \{(y,x)\ |(x,y) \in R_i\} = R_i$

$A_3$ the cardinal of $\{z \in X\ |\ (x,z) \in R_i$ et $(y,z) \in R_j\}$ is a constant $p_{ij}^k$ independent of the choice of (x,y) in $R_k$.

Let us now define the adjacency matrices indexed by X × X with entries in C ; for i = 0,1,...,t :

$$D_i = [D_i(x,y)] \text{ where } D_i(x,y) = \begin{cases} 1 & \text{if } x\ R_i\ y \\ 0 & \text{otherwise} \end{cases}$$

We can restate the preceding axioms in matrices terms

$A'_1$ $D_0 = I$

$A'_2$ for $i = 0, 1, \dots t$ $D_i^t = D_i$

$A'_3$ $D_i D_j = \sum_{k=0}^{t} p_{ij}^k D_k$

It can be shown [9], [7] that the $D_i$ generate and are a basis of a semi simple commutative algebra, called the Bosé Mesner algebra with a basis of t+1 idem potents $J_i$ s.t.

$$J_i J_k = \delta_{ik} J_i$$

$$\sum_{i=0}^{t+1} J_i = I$$

Then it is easy to see that there exists reals numbers $p_k(i)$ s.t.

$$D_k J_i = p_k(i) J_i$$

These $p_k(i)$ are called the "first eigenvalues" of the scheme.

The "second eigenvalues" $q_k(i)$ can be defined as follows :

$$J_k = \frac{1}{|X|} \sum_{i=0}^{t} q_k(i) D_i.$$

In matrix form, we have :

$$P_{i,k} = p_k(i) \; ; \; Q_{i,k} = q_k(i)$$

$$PQ = |X| I$$

In the Hamming scheme P is the matrix over the basis of monomials of the famous Mc Williams transform which maps weight enumerators of linear codes into those of their orthogonal dual [8], [13].

3. Definition of the Lee Scheme :

Fist, let us consider the scheme on $Z_q$ with $s = \left[\frac{q}{2}\right]$ classes (called ordinary q-gon in [3]) :

$$x R_k^1 y \Leftrightarrow x-y = \pm k.$$

For any vector z in $Z_q^n$ we can define its Lee composition, denoted by lc(z) :

$$lc(z) = (c_0, c_1, \dots, c_s),$$

where the $c_i$ are given by :

$$c_i = |\{j \in [0..n) \;/\; z_j = \pm i\}|.$$

We now define a scheme with $N = C_{n+s}^{s} - 1$ classes on $Z_q^n$ by :

$$x R_{\underline{k}} y \text{ iff } lc(x-y) = \underline{k}.$$

where $\underline{k}$ is any possible composition vector.

This scheme is called the "extension of order n" of the ordinary q-gon [9]. Obviously the Lee metric is constant on the classes of the scheme :

$$x\, R_k\, y \Rightarrow d_L(x,y) = \|k\| = k_1 + k_2 + \ldots + sk_s.$$

However the relations $R''_k$.

$$x\, R''_k\, y \text{ iff } d_L(x,y) = k$$

do <u>not</u> yield in general an association scheme, as can be seen by drawing a picture in the peculiar case n = 2, i = 2, j = 2, k = 4.

We shall denote this scheme by L(n,q) [16].

## 4. The Bose-Mesner algebra

If we denote by $D^L_{\underline{k}}$ (resp. $D^H_i$) the generic adjacency matrix of the Lee (resp. Hamming) scheme, we obtain the relationship

$$D^H_i = \sum_{|h|=i} D^L_k$$

where $|h| = h_1 + h_2 + \ldots + h_s$ denote the Hamming weight of any vector of Lee composition $\underline{h}$.

That means that the Lee scheme is a "refinement" of the Hamming scheme. We can continue the comparison in the same vein.

$$J^H_i = \sum_{|h|=i} J^L_k$$

$$P^H_i(|\ell|) = \sum_{|k|=i} P^L_k(\underline{\ell})$$

Where the $P^H_i(j)$ are the celebrated Krawtchouck polynomials.

Moreover, if we define the inner product of two vectors x and y of $Z^n_q$ in the following way

$$\langle x,y\rangle = \prod_{i=1}^{n} w^{x_i y_i}$$

where w is a $q^{\underline{th}}$ primitive root of unity over C, the following equalities hold :

$$J_{\underline{k}}(x,y) = \sum_{lc(z)=\underline{\ell}} \langle x-y,z\rangle$$

$$P_{\underline{k}}(\underline{i}) = \sum_{lc(y)=\underline{k}} \langle x-y\rangle \text{ for any } x \text{ with } \ell_c(x) = \underline{i}$$

These relations, which have analogous counterparts in the Hamming scheme, are a particular case of a phenomenon occuring for scheme (X,R) over an abelian group X, where the $R_i$ are invariant by the translation due to the group law [8 p.23].

By using properties of the inner product one can obtain a closed form for the generating function of the $p_{\underline{k}}(\underline{i})$, [2].

If we set :

$$P_{\underline{i}}(\underline{z}) = \sum_{\underline{k}} p_{\underline{k}}(\underline{i})\; z_1^{k_1} \ldots z_s^{k_s}$$

We obtain for q odd :

$$P_{\underline{i}}(\underline{z}) = \prod_{\ell=0}^{s} \Big(1 + 2 \sum_{m=1}^{s} z_m \cos\big(\frac{2\Pi}{q} m\ell\big)\Big)^{i_\ell}$$

and q even :

$$P_{\underline{i}}(\underline{z}) = \prod_{\ell=0}^{s} \left(1+2 \sum_{m=1}^{s-1} z_m \cos\left(\frac{2\Pi}{q} m\ell\right) + (-1)^{\ell} z_s\right)^{i_\ell}$$

All these relations and the links with Krawtchouk polynomials show that the $p_{\underline{k}}(\underline{i})$ are polynomials in s variables $i_1, i_2, \ldots, i_s$ of total degree at most $|k|$.

Moreover Bannai has shown [3] that in a polynomial scheme (i.e. where the $p_k(i)$ are polynomials in <u>one</u> variable) with sufficiently many classes the $p_k(i)$ are rational numbers. Hence, the Lee scheme is not, in general, polynomial, because of the cosinus terms.

## 5. The "Lloyd Theorem" in Lee metric :

In this section we show the link between Bassalygo's result and the Lee scheme. Consider the generalised Lloyd polynomial in s variables :

$$\psi_e(\underline{z}) = \sum_{\|k\| \le e} p_{\underline{k}}(\underline{z}).$$

We denote by $\pi_{q,n}(e)$ the number of distinct Lee compositions in the Lee sphere of radius e centered at the origin $V_{n,e}(q)$.

**Theorem** : If there exists a perfect e-error correcting code C in L(n,q) then $\psi_e(\underline{z})$ has at least $\Pi_{q,n}(e) - 1$ distinct roots.

**Proof** : We consider the matrices $\hat{D}_{\underline{i}}$ with integer entries, and size $(N+1)\times(N+1)$

$$\hat{D}_{\underline{i}}(\underline{j},\underline{k}) = p_{\underline{i},\underline{j}}^{\underline{k}}$$

It in well-known [5], [14] that these matrices form an algebra isomorphic to the Bose Mesner algebra and have the same left eigenvalues. In particular there is a basis which diagonalize all the $\hat{D}_{\underline{i}}$ with the <u>complete</u> system of eigenvalues :

$$p_{\underline{i}}(\underline{j}) \quad \underline{j} \text{ taking N+1 values.}$$

We call $\underline{a}$ the vector of the inner distribution of C with entries :

$$a_{\underline{i}} = \frac{|C \cap R_{\underline{i}}|}{|C|}$$

and $\underline{h}$ the columm vector with entries :

$$h_{\underline{i}} = \frac{|R_{\underline{i}}|}{q^n}$$

The relation :

$$(I + \sum_{\|i\| \le e} \hat{D}_i)\, \underline{a} = \underline{k}$$

expresses the perfection of C. This is a direct generalisation of a fact already observed in the context of distance transitive graphs [6].

Now the translated code C+h (h non zero) is also perfect with inner distribution $\underline{a}_h$. Thus we have that :

$$(I + \sum_{\|i\| \le e} \hat{D}_i)\, (\underline{a} - \underline{a}_h) = \underline{0}.$$

When h runs over $V_{n,e}(q)$ with distinct Lee compositions, this yields $\pi_{q,n}(e)-1$ distinct and linearly independent eigenvectors, and, consequently, as many null eigenvalues of the "Lloyd operator" of matrix :

$$(I + \sum_{\|i\| \le e} \hat{D}_i).$$ Q.E.D.

6. <u>An asymptotic estimate for</u> $\Pi_{q,n}(e)$ :

For convenience we set :

$$\delta_{q,n}(e) = \Pi_{q,n}(e) - \Pi_{q,n}(e-1),$$

which is equivalent to :

$$\delta_{q,n}(e) = |\{\underline{h} \in \mathbb{N}^s | \sum_{i=1}^{s} k_i \le n \text{ and } \sum_{i=1}^{s} i\, k_i = e\}|.$$

We recall that a partition of the integer e is a finite non decreasing sequence of positive integers $\lambda_1, \lambda_2, \ldots, \lambda_r$, called "parts" such that :

$$\sum_{i=1}^{n} \lambda_i = e$$

<u>example</u> : 4 has 4 partitions 4 = 1+1+1+1=1+2+1=1+3=2+2.

So $\delta_{q,n}(e)$ counts the number of partitions of e in at most n parts with values in [1..s] (taking $k_i$ as the number of occurences of i in the sum).

If we put :

$$(x)_n = (1-x^n)(1-x^{n-1})\ldots(1-x),$$

we can define the Gaussian binomial coefficient of basis x by :

$$\begin{bmatrix} n \\ m \end{bmatrix} = \frac{(x)_n}{(x)_m (x)_{n-m}} .$$

Then, it can be shown [1] that :

$$\sum_e \delta_{q,n}(0) \quad x^e = \begin{bmatrix} n+s \\ s \end{bmatrix} = \begin{bmatrix} n+s \\ n \end{bmatrix}$$

and, consequently :

$$\sum_e \Pi_{q,n}(e) \quad x^e = \frac{1}{1-x} \quad \begin{bmatrix} n+s \\ n \end{bmatrix}$$

If n≥e and s≥e each part is necessarily ≤ s and there cannot be more than n non zero parts. We have then for $\delta_{q,n}(e)$ the generating function of ordinary unrestricted partition :

$$\prod_{i=1}^{\infty} \frac{1}{1-x^i} .$$

Then a deep result due to Ramanujan [1] tells us that :

$$\delta_{q,n}(e) \underset{e\infty}{\sim} \frac{1}{4e\sqrt{3}} \exp [\Pi \sqrt{\frac{2e}{3}}].$$

Then, after some manipulations we obtain :

$$\Pi_{q,n}(e) \underset{e\infty}{\sim} \frac{1}{2\Pi\sqrt{2e}} \exp [\Pi \sqrt{\frac{2e}{3}}].$$

7. <u>The Lee compositions distribution of the translate of a code</u> :

Let $B_{\underline{i}}(e)$ be the Lee composition distribution of the translate C-e

$$B_{\underline{i}}(e) = |\{a \in C| \; a \; R_{\underline{i}} e\}|$$

Let $\beta^{\underline{m}}(\underline{x})$ be the annihilator polynomial of the $\underline{m}$ $\underline{th}$ order.

$$\beta^{\underline{m}}(\underline{x}) = \frac{q^n}{|C|} \; x^{\underline{m}} \prod_{\underline{i} \in J} \prod_{\ell=1}^{s} (1 - \frac{x_\ell}{i_\ell}).$$

J is the set of indices where the dual $\underline{a}' = P^T \underline{a}$ of the inner distribution, $\underline{a}$ of the code C is nonzero. In case of C linear $\frac{1}{|C|} P^T \underline{a}$ is the inner distribution of the dual of C [2]. Let $\beta^{\underline{m}}_{\underline{i}}$ be the development of the polynomial $\beta^m(\underline{x})$ in the basis of the $P_{\underline{i}}(\underline{x})$.

We denote by $\underline{0}$ the composition of the null vector.

Theorem : The inner distribution of C-e can be obtained by the linear recurrence :

$$\sum_{\underline{i}\in J} \beta_{\underline{i}}^{m} B_{\underline{i}} = \delta_{\underline{0},\underline{m}} \quad \text{(Kronecker symbol)}$$

Proof : We shall admit the following result which is a straightforward generalisation from Hamming [8] to Lee scheme :

$$\sum_{e\in Z_q^n} (B'_{\underline{k}}(e))^2 = q^n |C| a'_{\underline{k}} \text{ for any } \underline{k} \neq 0$$

where $B'_k(e)$ is the dual transform of $B_{\underline{k}}(e)$, i.e :

$$B'_k(e) = \sum_{\underline{i}} p_{\underline{k}}(\underline{i}) B_{\underline{i}}(e)$$

From the very definition of J we have that :

$$\beta^m(\underline{k}) a'_{\underline{k}} = 0 \text{ for } \underline{k} \neq \underline{0}$$

Using the cited result this implies :

$$\beta^m(\underline{k}) B'_{\underline{k}}(e) = 0 \text{ for } k \neq \underline{0}$$

We need the following identity :

$$\sum_{\underline{i}} \alpha_{\underline{i}} A_i = \frac{1}{q^n} \sum_{L} \alpha(\underline{k}) A'_k \text{ where } \alpha(\underline{x}) = \sum_{\underline{i}} \alpha_i P_{\underline{i}}(x)$$

which comes simply from the fact that $P^2 = q^n I$ and

$$\langle\underline{\alpha}, \underline{A}\rangle = \frac{1}{q^n} \langle P_{\underline{\alpha}}, P^T \underline{A}\rangle$$

where <,> denotes the ordinary inner product on $\mathbb{R}^{N+1}$.

Applying this identity, with $\alpha = \beta^{\underline{m}}$ and $\underline{A} = (B_{\underline{i}}(e))_{\underline{i}}$ yields :

$$\sum_{\underline{i}} \beta_{\underline{i}}^{m} B_{\underline{i}}(e) = \frac{1}{q^n} \beta(\underline{0}) B'_{\underline{0}}(e).$$

Since $p_{\underline{0}}(\underline{x}) = 1$ we have that $B'_{\underline{0}}(e) = |C|$. Q.E.D.

## 8. An upper bound on covering radius :

We denote by $\delta'$ the integer :

$$\delta' = \text{Max } \{\|i\| \,|\, \underline{i} \in J\}.$$

with the same notations as in the preceding paragraph. If C is linear $\delta'$ is the diameter of its dual.
We define the covering radius of C :

$$\rho = \operatorname*{Max}_{e \notin C} d_L(e,C).$$

Theorem : $\rho \leq \delta'$

Proof : We use the recurrence on the distribution with $\underline{m} = \underline{0}$, so that $\delta_{\underline{0},\underline{m}} = 1$.

Now, all the summands in the recurrence cannot vanish together. $B_{\underline{i}}(e) \neq 0$ means that there exists an a with

so that : $d_L(a,e) = \|i\|$

$\rho \leq \|i\| \leq \delta'$ Q.E.D.

example : We take the perfect negacyclic code of length 2 over $Z_{13}$ with generator polynomial : $g(x) = x+5$ [5]. It is self dual with "negacycle" representatives : (0,0) ; (1,5) ; (3,2) ; (6,4) we have $\rho = 2$ and $\delta' = 10$.

We remark that the number of distinct Lee composition of the dual is 3, which could not happen in Hamming scheme where $s' = e$ for a perfect code [8].

Remark : In fact a stronger result holds [15]. Let $s'$ be the number of distinct nonzero coefficient $a'_k$ (for $k \neq \underline{0}$) in the dual inner distribution of C then

$$\rho \leq s'$$

## 9. T-designs :

A T-design in a general association scheme (X,R) is a subset Y of X, such that :

$$\forall_i \in T, \quad a'_i\,(Y) = 0$$

where $a'_i$ is the dual inner distribution.

It has been shown [8] that T-designs in the Hamming scheme for $T = \{1, 2, \ldots t\}$ are orthogonal arrays of strength t, with n constraints, q level and index $|Y| \,|\, q^t$.

Let $T^H_\tau$ denote the set of indices $T^H_\tau = \{\underline{i} \,|i| \leq \tau\}$.

Proposition : a $T^H_\tau$-design in the Lee scheme is exactly a $\{1,\ldots,\tau\}$ design in Hamming scheme.

Proof : Let A' (resp-a') denote the dual inner distribution in the Hamming (resp. Lee) scheme.

As in paragraph 4, by using the fact that the Lee scheme is a refinement of the Hamming scheme, it can be proved that :

$$A'_j = \sum_{|i|=j} a'_{\underline{i}}$$

The proposition follows by the nonnegativity of the $A'_j$ and the $a'_{\underline{i}}$. Q.E.D.

Using the obvious bound

$$d_L \leq s.d_H$$

we see that :

$$T^L_\tau \supset T^H_{[\tau/s]}$$

so that any T-design in L(n,q) is a particular orthogonal array in H(n,q).

## 10. Characterization of T-designs :

We define a generalized character for a subset Y and any u in $Z_q^n$ :

$$\chi_u(Y) = \sum_{v \in Y} \langle u,v \rangle$$

Proposition : Y is a T-design iff ($\ell c(u) \in T \Rightarrow \chi_u(Y) = 0$)

Proof : We introduce the characteristic matrices of the code C in the following way :

$$H_{\underline{k}} = [\langle a,h \rangle] \; a \in C \quad \ell c(h) = \underline{k}$$

Using the expression of $p_{\underline{k}}(\underline{C})$ in function of inner products, we obtain :

$$H_{\underline{k}} \, \bar{H}^T_{\underline{k}} = [P_{\underline{k}}(\ell_c(a-b)) \; a,b \in C].$$

Let $\underline{j}$ be the all one vector of size $|C|$.

$$\| \underline{j} \, H_{\underline{k}} \|^2 = |C| \, a'_{\underline{k}} = \sum_{\ell c(u)=\underline{k}} |\chi_u(Y)|^2.$$

So $a'_{\underline{k}} = 0$ iff $\chi_u(Y) = 0$ for any u such that $\ell c(u)=\underline{k}$ Q.E.D.

Example : We write the words of the same self dual perfect code in columns :

| | | | | | | | | | | | | |
|---|---|---|---|---|---|---|---|---|---|---|---|---|
| 1 | 2 | 3 | 4 | 5 | 6 | 0 | -1 | -2 | -3 | -4 | -5 | -6 |
| -5 | 3 | -2 | 6 | 1 | -4 | 0 | 5 | -3 | 2 | -6 | -1 | 4 |

It is, from [9], an orthogonal array of strength one ; each row contain each symbol of $Z_{13}$ exactly once.

Taking u = (1,-1) in the preceding theorem and writing the row of differences yields :

6 -1 5 -2 4 -3 0 -6 1 -5 2 -4 3

Each symbol of $Z_{13}$ is seen exactly once.

The same result would occur with any u such that

$$W_L(u) \leq 4.$$

We see that the Lee metric yields more information on the same object than the Hamming one.

## 11. Open problems and conclusion :

The Lee scheme bears many ressemblances with the Hamming scheme. The underlying set is an abelian group allowing us , as noticed in [9 p.23], to express the first eigenvalues in terms of inner products. In particular, this led to a closed form for the generating function of these eigenvalues, and an algebraic characterisation for the T-designs. A combinatorial characterization would depend on the size of the alphabet as can be seen by examples.

However there is a main difference : the Hamming scheme is both P and Q polynomial [9] and the Lee scheme is not, in general, as noticed in §4. We have to manipulate polynomials with irrational coefficients in many variables, instead of polynomials with rational coefficients in one variable. In particular the number of zeroes over a compact set is not easy to evaluate.

This is the reason why the Lee version of Lloyd theorem is so complicate and hard to apply. Even if we had more accurate estimates of $\pi_{q,n}(e)$, bounding the number of lattice points solution of the Lloyd equation would still remain a difficult arithmetical problem.

12. Bibliography :

[1] G. ANDREWS "The theory of partitions" Addison Wesley Encyclopedia of Math and its applications.

[2] J. ASTOLA "The theory of Lee codes" Research report 1982 University of Lappeenranta, Finland.

[3] E. BANNAI and T. ITO "Algebraic combinatorics". Benjamin Cummings.

[4] L.A. BASSALYGO "A necessary condition for the existence of perfect codes in the Lee-metric" Mat. Zametki 15, p. 313-320, 1974.

[5] E.R. BERLEKAMP "Algebraic coding theory" Revised 1984 edition Aegean Park Press.

[6] N. BIGGS "Finite groups of automorphisms". Cambridge University Press.

[7] R.C. BOSE and D.M. MESNER (1959) "On linear associative algebras corresponding to association schemes of partially balanced designs" Ann. Math. Statist. 10 p. 21-38.

[8] P. DELSARTE "Four fundamental parameters of a code and their combinarorial significance" Info and Control 1973, 23, P. 407-438.

[9] P. DELSARTE "An algebraic approach to the association schemes of coding theory" Philips Res. Rpts. Suppl 1973 N° 10.

[10] S.W. GOLOMB and L.R. WELCH "Perfect codes in the Lee metric and the packing of polyominoes" SIAM J. Appl. Math. Vol 18 N° 2 pp. 302-317, 1970.

[11] S.W. GOLOMB and L.R. WELCH "Algebraic coding and the Lee metric" in Error correcting codes. H.B. MANN ed. p. 175-194 Wiley N.Y. 1968.

[12] C.Y. LEE 'Some properties of non binary error correcting codes" IEEE transactions on Information Theory. IT-4 p. 77-82.

[13] F.J. MAC WILLIAMS (1961) Doctoral Disertation Harvard University unpublished.

[14] F.J. MAC WILLIAMS and N.J.A. SLOANE "The theory of error correcting codes" North Holland.

[15] P. SOLE "On the parameters of Lee Codes" in preparation.

[16] H. TARNANEN "An approach to constant weight and Lee codes by using the methods of association schemes" Thesis Turku University 1982. Finland.

# ON MODULAR WEIGHTS IN ARITHMETIC CODES

# AU SUJET DES POIDS MODULAIRES DANS LES CODES ARITHMETIQUES

Antoine LOBSTEIN
Centre National de la Recherche Scientifique UA 820
Ecole Nationale Supérieure des Télécommunications
46 rue Barrault, 75634 PARIS Cedex 13, FRANCE

*Résumé:* Les codes arithmétiques sont utilisés pour détecter et corriger des erreurs survenant lors d'additions, modulo un entier M strictement positif, effectuées sur des entiers. Pour décrire de manière adéquate le poids de telles erreurs, Garcia et Rao ont introduit la notion de distance modulaire entre entiers (relative à un modulo M>0 et une base r>1), qui présente l'inconvénient de ne pas toujours satisfaire l'inégalité triangulaire (cependant, celle-ci est vérifiée dans les cas le plus souvent utilisés en pratique). Pour y remédier, Clark et Liang ont donné une nouvelle définition de la distance modulaire, de manière que l'inégalité triangulaire soit satisfaite dans tous les cas.
Nous examinons le problème de déterminer les valeurs de M pour lesquelles ces deux définitions coïncident.

*Abstract:* Errors arising in addition modulo M in computer computations can be detected and corrected by arithmetic codes. To more adequately describe the weight of such errors, Garcia and Rao introduced the modular distance (relative to modulus M>0 and radix r>1) which, unfortunately, does not, in general, satisfy the triangular inequality (however, it does hold for the cases of greatest practical interest). To supply it, Clark and Liang gave a new definition of modular distance, which always satisfies the triangular inequality.
We investigate the problem of when these two definitions are identical.

*Mots clés:* Codes arithmétiques, forme de base r, forme minimale, forme modifiée de base r, forme non-adjacente généralisée, inégalité triangulaire, poids arithmétique, poids modulaire.

*Key words:* Arithmetic codes, arithmetic weight, generalized nonadjacent form, minimal form, modified radix-r form, modular weight, radix-r form, triangular inequality.

This work was supported by a grant from University of Technology of Eindhoven (Netherlands).

## NOTATIONS

We shall use the following notations:
$|A|$ is the number of elements of a nonempty set A;
For a,b in Z: $[\![a,b]\!]=[a,b]\cap Z$;
For a in $Z^*$, b in Z: $a|b$ means that a divides b;
For a,b in Z, M in $N^*$: $a\equiv b$ [M] means that $M|(a-b)$.

## 1. INTRODUCTION

*Arithmetic codes* have been designed in order to detect and correct errors arising when adding (modulo an integer $M>0$) integers represented in a radix $r>1$. This is achieved by incorporating *redundancy* inside the integers: for instance, in an AN-code, an integer is encoded by multiplying it by a fixed integer, A. Thus the sum of two codewords is a multiple of A.

We now give some definitions; from now on, r is an integer greater than, or equal to, 2, *fixed*.

Any positive (respectively negative) integer N has a *unique* representation of the form $N=\sum_{i=0}^{n} a_i r^i$ , where $a_i \in Z$ and $0\leqslant a_i<r$ (respectively $-r<a_i\leqslant 0$) for all $i=0,1,\ldots,n$. Such a representation is referred to as the *radix-r form* for N.

DEFINITION 1.1 A *modified radix-r form* of an integer N is

$$N=\sum_{i=0}^{n} e_i r^i \qquad [1.1]$$

where $e_i \in Z$ and $|e_i|<r$ for all $i=0,1,\ldots,n$.

This form is not unique.

DEFINITION 1.2 A form [1.1] is called *minimal* if the number of its nonzero coefficients is minimal.

DEFINITION 1.3 The *arithmetic weight*, W(N), of an integer N is the number of nonzero coefficients in a minimal form of N.

A minimal form is not unique. However, if the coefficients $e_i$ $(i=0,1,$

...,n-1) in [1.1] satisfy the condition

$$\left.\begin{array}{l} e_i.e_{i+1}=0 \\ \text{or } (e_i.e_{i+1}>0 \text{ and } |e_i+e_{i+1}|<r) \\ \text{or } (e_i.e_{i+1}<0 \text{ and } |e_i|<|e_{i+1}|) \end{array}\right\} \quad [1.2]$$

then form [1.1] is *unique and minimal*, and is referred to as the *generalized nonadjacent form (GNAF)* for N.

Such a form exists for all integers (for more details about the above basic definitions and properties, see for instance [1] or [3]).

<u>*Example.*</u> *r=10, N=1882.*
$N=1.10^3+8.10^2+8.10^1+2.10^0$ *(a)*, $N=1.10^3+8.10^2+9.10^1-8.10^0$ *(b)*,
$N=2.10^3-1.10^2-1.10^1-8.10^0$ *(c)*, $N=1.10^4-9.10^3+8.10^2+8.10^1+2.10^0$.
*(a) is the radix-10 form of 1882; (c) is the GNAF of 1882; (a), (b) and (c) are minimal, and W(1882)=4 (with respect to radix 10).*

There are simple algorithms determining the GNAF (hence the arithmetic weight) of an integer, starting from its radix-r form, or from any modified radix-r form [3].

The arithmetic weight is appropriate to the measure of errors arising in additions of integers: if an erroneous sum differs from the exact sum by a quantity E, it is natural to see the weight of error E as the least number of terms $e_i r^i$ ($|e_i|<r$) with sum E.

Let us now consider the integers modulo M ($M \in N^*$). For $I_1, I_2$ in $[\![0,M-1]\!]$, we write

$I_1 \oplus I_2 = \begin{cases} I_1+I_2 & \text{if } I_1+I_2<M \\ I_1+I_2-M & \text{otherwise} \end{cases}$ ; if I is the result of an addition with exact result C, we define the ring error, F, by setting $I=C\oplus F$. The error E, given by $I=C+E$, is equal, according to its sign, to F or F-M. The arithmetic weight of E being W(F) or W(F-M) (=W(M-F)), we are led to the following definition, introduced by Rao and Garcia [1]:

<u>DEFINITION 1.4</u> Let $M \in N^*$; for every a in $[\![0,M-1]\!]$, the *modular weight*, $w_M^*(a)$, of a is the minimum of W(a) and W(M-a).

This definition of modular weight does <u>not</u>, in general, satisfy the triangular inequality. However, it does hold for the cases of greatest practical interest, namely $M=2^n$ or $M=2^n\pm1$.

Clark and Liang [4] gave a new definition of modular weight, which, from a mathematical viewpoint, has the advantage that it always satisfies the triangular inequality:

<u>DEFINITION 1.5</u> Let $M \in N^*$; for every a in Z, the *modular weight*, $w_M(a)$, of a is given by $w_M(a) = \text{Min}(W(X)/X \equiv a\ [M])$.

<u>*Examples.*</u> *(a)* $r=2$, $M=251=2^8-2^2-2^0$.

$N_1=64=2^6$; $M-N_1=187=2^8-2^6-2^2-2^0$; $w_M^*(N_1)=1$.

$N_2=240=2^8-2^4$; $M-N_2=11=2^4-2^2-2^0$; $w_M^*(N_2)=2$.

$N_1 \oplus N_2=53=2^6-2^4+2^2+2^0$; $M-(N_1 \oplus N_2)=198=2^8-2^6+2^3-2^1$;

$w_M^*(N_1 \oplus N_2)=4>w_M^*(N_1)+w_M^*(N_2)$.

*(b)* $r=2$, $M=11=2^4-2^2-2^0$.

$w_M^*$ *satisfies the triangular inequality.*

*But* $w_M^*(5)=2$ *(because* $W(5)=W(6)=2$*), and* $w_M(5)=1$ *(because* $W(16)=1$*).*

$w_M^*$, being derived by comparing two arithmetic weights, is easy to compute. As far as we know, nothing has been said about the complexity of computing the modular weight $w_M$.

Clark and Liang noticed that, if $M=r^n, r^n-1$ or $r^n+1$, then for every a in $[\![0,M-1]\!]$, one has $w_M(a)=w_M^*(a)$.

Two questions arise:

1) For which M is the triangular inequality satisfied by $w_M^*$?

2) For which M are the two definitions of modular weight equivalent? (in the sense that, $\forall a \in [\![0,M-1]\!]$, $w_M(a)=w_M^*(a)$).

Of course, any M for which question 2) gets a positive answer is such that question 1) gets a positive answer; so the only cases for which we shall have to determine whether the two definitions are equivalent or not are the cases that satisfy the triangular inequality. (the above example (b) already shows that it is possible that $w_M^*$ satisfies the triangular inequality without the two definitions being equivalent)

Now question 1) has got an exhaustive answer by Ernvall [5] [6] [7], and hence we shall make a large use of her results, which we give in Appendix.

## 2. PRELIMINARIES

In this section, Theorem 2.4 gives *sufficient, and easy to check,* conditions for the equivalence of the two definitions of modular weights.

The *general idea* is to replace, in a modified radix-r form of any integer Y, each term of "high" degree by a term (congruent modulo M) of "low" degree, in order to "carry" Y inside the interval $[\![-M+1,M-1]\!]$:

```
                           -M+1                      M-1
--------+---•------------+------+---------------+----------->
        Y---•-------------------> Y' (Y≡Y' [M])
```

LEMMA 2.1 Let $M\in\mathbb{N}^*$; let Y be any integer, represented by a modified radix-r form: $Y=\sum_{i=0}^{s-1} b_i r^i$, where $|b_i|<r$ for $i\in[\![0,s-1]\!]$. Let $S=|\{b_i\neq 0 \ / \ i\in[\![0,s-1]\!]\}|$. If there is an integer Y' satisfying

$Y'\equiv Y\ [M]$ [2.1.1]

$|Y'|<M$ [2.1.2]

and Y' can be represented by a modified radix-r form with at most S nonzero coefficients [2.1.3]

then $\forall X\in[\![0,M-1]\!]$, $w_M(X)=w_M^*(X)$.

Proof. Let X be in $[\![0,M-1]\!]$. Because $W(M-X)=W(X-M)$, we get from the two definitions of modular weight that $w_M(X)\leq w_M^*(X)$. Let Y be an integer reaching the (Clark and Liang) modular weight of X: $w_M(X)=W(Y)$, and $Y\equiv X\ [M]$. We now assume that Y is given by a minimal form: $W(Y)=S$. If Y', satisfying the three conditions of Lemma 2.1, exists, then, since $Y'\equiv X\ [M]$ and $|Y'|<M$, necessarily $Y'=X$ or $Y'=X-M$, and so $w_M^*(X)\leq W(Y')$. So, because $W(Y')\leq S$, $w_M^*(X)\leq w_M(X)$, and finally $w_M(X)=w_M^*(X)$. ◊

Lemma 2.2 states that, in order to show that each term of "high" degree can be replaced by a term (congruent modulo M) of "low" degree, it is sufficient to check this property on a set of r-1 terms.

LEMMA 2.2 Let $M\in\mathbb{N}^*$, represented by its GNAF: $M=ar^n+\sum_{i=0}^{n-1} a_i r^i$, where $0<a<r$, and, for all $i\in[\![0,n-1]\!]$, $|a_i|<r$ and condition [1.2] is satisfied (with $a_n=a$). Let H be defined as follows:

if $a=1$, $H=\{r^n,2r^n,\ldots,(r-1)r^n\}$, and, if $a>1$,

$H=\{ar^n,(a+1)r^n,\ldots,(r-1)r^n,r^{n+1},\ldots,(a-1)r^{n+1}\}$.

If $\forall L\in H$, $\exists B\in\mathbb{Z}$, $j\in\mathbb{N}$ such that

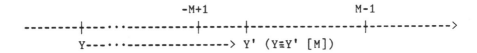

$Br^j \equiv L$ [M]

$|B|<r$ [2.2.1]

and $(j<n)$ or $(j=n$ and $|B|<a)$ [2.2.2]

then $\forall t \in N^*$, $\forall e \in [\![-r+1,r-1]\!]$, $\exists B' \in Z$, $j' \in N$ satisfying $B'r^{j'} \equiv er^{n+t}$ [M], [2.2.1] and [2.2.2].

Proof. First, notice that if L is the opposite of an element of H, for which suitable B and j have been found, then $B'=-B$ and $j'=j$ are suitable for L. The proof is by induction on t.

Case t=1. Let $Y=er^{n+1}$; we assume, without loss of generality, that $0<e<r$. If $e \leqslant a-1$, $Y \in H$; so we assume $e>a-1$. Now $er^n \in H$: $\exists B_o$, $j_o$ verifying $B_o r^{j_o} \equiv er^n$ [M], [2.2.1] and [2.2.2]. And $Y \equiv B_o r^{j_o+1}$ [M].

Case 1. $j_o<n-1$. Choose $B'=B_o$, $j'=j_o+1$.

Case 2. $j_o=n-1$. Then $Y \equiv B_o r^n$ [M]. If $|B_o|<a$, choose $B'=B_o$ and $j'=n$. If $|B_o| \geqslant a$, either $B_o r^n$ or $-B_o r^n$ belongs to H.

Case 3. $j_o=n$. $Y \equiv B_o r^{n+1}$ [M], with $|B_o|<a$. Then either $B_o r^{n+1}$ or $-B_o r^{n+1}$ belongs to H.

Induction assumption. $\forall d \in [\![-r+1,r-1]\!]$, $\exists B$, j satisfying $Br^j \equiv dr^{n+t-1}$ [M], as well as [2.2.1] and [2.2.2] (for $t>1$).

Let $Y=er^{n+t}$, with $0<e<r$. Using the induction assumption, $\exists B_o$ and $j_o$ verifying $Y \equiv B_o r^{j_o+1}$ [M], [2.2.1] and [2.2.2]. The analysis of the cases is the same as in the case t=1. ◊

Lemma 2.3 states that, in a modified radix-r form, replacing a term by another one cannot increase the number of nonzero terms.

LEMMA 2.3 Let Y be an integer represented by a modified radix-r form: $Y=\sum_{i=0}^{s-1} b_i r^i$, where $|b_i|<r$ for $i \in [\![0,s-1]\!]$. Let $S=|\{b_i \neq 0 \;/\; i \in [\![0,s-1]\!]\}|$. Let $Y_1=Y+c_{j_o} r^{j_o}$ $(0<|c_{j_o}|<r)$.

Then $Y_1$ can be represented by a modified radix-r form containing at most S+1 nonzero terms.

The proof is easy [8].

Now we can give Theorem 2.4, which shows that, with the exception of some cases (due to technical conditions about M), to find B and j for each element in H is sufficient to show the equivalence of modular weights.

The exclusion of some forms of M will be no trouble, because we shall see that they are forms for which the modular weights are not equivalent.

Finally, let us point out that in the only case W(M)=3 that we shall have to deal with, we shall need slightly different sufficient conditions to prove the equivalence of modular weights. For sake of simplicity, we do not give them here.

<u>THEOREM 2.4</u> Let $M\in N^*$, represented by its GNAF: $M=ar^n+\sum_{i=0}^{n-1} a_i r^i$, where $0<a<r$, and, for all $i\in[\![0,n-1]\!]$, $|a_i|<r$ and condition [1.2] is satisfied (with $a_n=a$). Let H be defined as in Lemma 2.2. If $(M\geq ar^n)$, or $(M<ar^n$ and $a\neq 1)$, or $(M<ar^n$, $a=1$ and $W(M)=2)$, and if $\forall L\in H$, $\exists B\in Z$, $j\in N$ satisfying $Br^j\equiv L$ [M], [2.2.1] and [2.2.2],
then $\forall X\in[\![0,M-1]\!]$, $w_M(X)=w_M^*(X)$.

<u>Proof.</u> Let Y be any integer, represented by a modified radix-r form: $Y=\sum_{i=0}^{s-1} b_i r^i$, where $|b_i|<r$ for $i\in[\![0,s-1]\!]$. We assume that Y is positive. For Y, we want to find an integer Y' verifying the conditions [2.1.1], [2.1.2] and [2.1.3] of Lemma 2.1.
<u>Case 1.</u> $M\geq ar^n$. If $s-1<n$, or if $(s-1=n$ and $b_{s-1}=b_n<a)$, then $Y<M$. In this case, $Y'=Y$ is suitable. Therefore we assume that either Y contains a term $b_n r^n$ where $b_n\geq a$, or Y contains terms of degree more than n. The hypothesis of Theorem 2.4 enables us to use Lemma 2.2, and state that, for any term in Y with degree more than or equal to n, say $b_t r^t$, we can find B and j satisfying $Br^j\equiv b_t r^t$ [M], [2.2.1] and [2.2.2]. So, starting from Y, we construct integers $Y_1,Y_2,\ldots$ by successively replacing the terms with degree more than n, or equal to n and having a coefficient superior or equal to a (in absolute value), by terms with degree less than n, or equal to n and having a coefficient inferior to a (in absolute value). By Lemma 2.3, none of these steps increases the number of nonzero terms. At the end, we get an integer $Y_F$ which is congruent to Y modulo M, and which can be represented by a modified radix-r form containing at most as many nonzero coefficients as the form of Y we started from. Moreover, this form contains no term with degree more than n, and either it contains no term with degree n, or it contains a term with degree n and coefficient inferior to a (in absolute value). So $|Y_F|<M$, and we choose $Y'=Y_F$.
<u>Case 2.</u> $M<ar^n$. We proceed in the same way as in Case 1, till we get $Y_F$. But now we cannot conclude that $|Y_F|<M$. If this is the case, then we choose $Y'=Y_F$. So we assume (without loss of generality) that $Y_F>M$, with $Y_F=\sum_{i=0}^{n} c_i r^i$, where $c_n<a$ and $|c_i|<r$ for $i\in[\![0,n-1]\!]$. Because M is given by its GNAF, we have $M\geq ar^n-(a-1)r^{n-1}-(r-a)r^{n-2}-(a-1)r^{n-3}-\ldots$
<u>Subcase 1.</u> $a>1$. Because $Y_F>M$, necessarily $c_n=a-1$ and $c_{n-1}\geq r-a>0$. Then $Y_F=(a-1)r^n+c_{n-1}r^{n-1}+\sum_{i=0}^{n-2} c_i r^i=ar^n+(c_{n-1}-r)r^{n-1}+\sum_{i=0}^{n-2} c_i r^i$, with $0>c_{n-1}-r>-r$. We have not increased the number of nonzero terms in the form of $Y_F$. We next replace $ar^n$, and get a suitable integer, Y'.
<u>Subcase 2.</u> $a=1$ and $W(M)=2$. Then $M\geq r^n-(r-1)r^{n-2}$, and $Y_F=\sum_{i=0}^{n-1} c_i r^i$. Because $Y_F>M$, necessarily $c_{n-1}=r-1$ and $c_{n-2}>0$. So $Y_F=(r-1)r^{n-1}+c_{n-2}r^{n-2}+\sum_{i=0}^{n-3} c_i r^i=r^n+(c_{n-2}-r)r^{n-2}+\sum_{i=0}^{n-3} c_i r^i$, we replace $r^n$, and we get Y'. ◊

*Example.* $M=2r^n-r^{n-1}$ $(r>2)$.

$H=\{2r^n, 3r^n, \ldots, (r-1)r^n, r^{n+1}=rr^n\}$.

$2r^n \equiv r^{n-1}$ $[M]$. Let $e \in [\![3,r]\!]$; we divide $e$ by 2: $e=2Q_o+R_o$, with $Q_o>0$ and $R_o=0$ or 1. If $R_o=0$, $er^n \equiv Q_o r^{n-1}$ $[M]$, with $Q_o<r$. If $R_o=1$, $er^n \equiv (Q_o+1-r)r^{n-1}$ $[M]$, with $|Q_o+1-r|<r$. So we are in the conditions of Theorem 2.4.

Assume $r$ is even and not a multiple of 4.

Let us consider $Y=2r^{n+3}+r^{n+1}$;

$r^{n+1} \equiv (\frac{1}{2}r)r^{n-1}$ $[M]$;

$2r^{n+3} \equiv r^{n+2}$ $[M] \equiv r(\frac{1}{2}rr^{n-1})$ $[M] \equiv (\frac{1}{2}r)r^n$ $[M] \equiv (\frac{1}{4}(r+2)-r)r^{n-1}$ $[M]$;

Hence $Y \equiv -\frac{1}{4}(r-2)r^{n-1}$ $[M]$: $Y'=-\frac{1}{4}(r-2)r^{n-1}$ satisfies, with respect to $Y$, the three conditions of Lemma 2.1.

All we have to do now is to repeatedly use Theorem 2.4 (mainly when $W(M)=2$) to determine when the modular weights are equivalent, the contrary being showed by counterexamples (existence of an integer X in $[\![0,M-1]\!]$ such that $w_M^*(X)>w_M(X)$).

Because of their extremely technical and repetitive character, from now on, we shall give no proof (except for the binary case), and the interested reader is referred to [8].

## 3. CASE r=2

Following Ernvall, in the binary case, the triangular inequality holds if and only if:

$W(M)<3$, or

$W(M)=3$ and the GNAF of M is of the form: $M=2^n+2^{n-2}+\varepsilon 2^i$ $(i<n-3)$ or $M=2^n-2^j+\varepsilon 2^i$ $(n-6<j<n-1,\ i<j-1)$, where $\varepsilon \in \{-1,1\}$.

THEOREM 3.1 In the binary case, the two modular weights are equivalent if and only if $W(M)<3$.

Proof. (a) If $W(M)<3$, using Theorem 2.4, it is sufficient to check that $2^n \equiv 0$ $[M]$, $2^n \equiv 2^j$ $[M]$ or $2^n \equiv -2^j$ $[M]$ $(j<n)$, which is obviously the case if $M=2^n$ or $M=2^n \pm 2^j$ $(j<n-1)$.

(b) $M=2^n+2^{n-2}+\varepsilon 2^i$; choose $X=2^n-2^{n-2}-\varepsilon 2^i$: $X \in [\![0,M-1]\!]$. $M+X=2^{n+1}$, so $w_M(X)=1$. $M-X=2^{n-1}+\varepsilon 2^{i+1}$. X and M-X are given here by their GNAFs, which shows that $w_M^*(X)=2>w_M(X)$.

(c) $M=2^n-2^j+\varepsilon 2^i$; choose $X=2^j-\varepsilon 2^i$: $X \in [\![0,M-1]\!]$, and is given by its GNAF. $M+X=2^n$, so $w_M(X)=1$. $M-X=2^n-2^{j+1}+\varepsilon 2^{i+1}$; this form is the GNAF of M-X, unless

$j+1=n-1$, in which case $M-X=2^{j+1}+\varepsilon 2^{i+1}$. This leads to $w^*_M(X)=2$. ◊

## 4. GENERAL CASE

### 4.1. W(M)=1.

THEOREM 4.1.1 If $W(M)=1$, then the two modular weights are equivalent.

### 4.2. W(M)=2.

THEOREM 4.2.1 If $W(M)=2$, the two modular weights are equivalent if and only if the GNAF of M has one of the following forms: ($a>0$, $b>0$, $\varepsilon\in\{-1,1\}$)
$M=2r^n-r^{n-1}$; $M=ar^n+\varepsilon\frac{1}{2}rr^{n-1}$ ($a>2$); $M=r^n+br^{n-1}$; $M=2r^n+2r^{n-1}$; $M=2r^n+br^{n-1}$ ($b|r$); $M=r^n+\varepsilon r^j$ ($j<n-1$); $M=r^n+\varepsilon\frac{1}{2}rr^{n-2}$; $M=2.4^n-2.4^{n-2}$; $M=3^n-2.3^{n-2}$.

### 4.3. W(M)=3.

LEMMA 4.3.1 If $W(M)=3$, the only case where the two modular weights *could be* equivalent is $M=r^n+br^{n-1}+\frac{1}{2}rr^{n-2}$ ($0<b<\frac{1}{2}r$, r even, $r>3$).

This form of the GNAF of M is one subcase of one of the twenty-two cases, verifying the triangular inequality, given by Ernvall (see Appendix).

LEMMA 4.3.2 If $W(M)=3$, and if the GNAF of M is $M=r^n+br^{n-1}+\frac{1}{2}rr^{n-2}$ ($0<b<\frac{1}{2}r$, r even, $r>3$),
if $r=4$ ($b=1$ is the only case), then the two modular weights are equivalent;
if $r=6$, if $b=1$ (resp. $b=2$), then the two modular weights are (resp. are not) equivalent;
if $r=8$, if $b=1$ or 2 (resp. $b=3$), then the two modular weights are (resp. are not) equivalent;
if $r=10$, if $b=1$ or 2 (resp. $b=3$ or 4), then the two modular weights are (resp. are not) equivalent;
if $r=12$, if $b=1$ (resp. $b=2,3,4$ or 5), then the two modular weights are (resp. are not) equivalent;
if $r\geq 14$, if $b=1$ (resp. $b=\frac{1}{2}r-1$), then the two modular weights are (resp. are not) equivalent.

For $r \geqslant 14$ and $1<b<\frac{1}{2}r-1$, *in most cases* the two modular weights are not equivalent, *in a few cases* they are equivalent, but unfortunately we have not managed yet to get a complete result.

### 4.4. $W(M)=4$.

THEOREM 4.4.1 If $W(M)=4$, then the two modular weights are not equivalent.

## 5. CONCLUSION

We investigated the problem of when the modular weight defined by Rao and Garcia, on the one hand, and the modular weight defined by Clark and Liang, on the other hand, are identical [Question 2)]. Our starting point was the knowledge of when the modular weight defined by Rao and Garcia satisfies the triangular inequality [Question 1); see complete answer by Ernvall in Appendix].

The following table summarizes our results:

| | $W(M)=1$ | $W(M)=2$ | $W(M)=3$ | $W(M)=4$ | $W(M)>4$ |
|---|---|---|---|---|---|
| Question 1) | YES | YES | YES in only 22 cases (see Appendix) | YES in only 10 cases (see Appendix) | NO |
| Question 2) | YES | YES in only 9 cases | -NO in 21 out of 22 cases, and most cases of the remaining case<br>-YES in some cases of the remaining case<br>- ? in the remaining cases | NO | NO |

Up to $r=13$, our results are complete. In particular, in the binary case, we have the following results:

| $r=2$ | $W(M)=1$ | $W(M)=2$ | $W(M)=3$ | $W(M)>3$ |
|---|---|---|---|---|
| Question 1) | YES | YES | YES in only 2 cases (see Section 3) | NO |
| Question 2) | YES | YES | NO | NO |

We see that, when studying a problem related to modular weight, it is necessary to clearly specify which definition of modular weight is used: the use of one or the other can lead to different conclusions [9].

## APPENDIX

Ernvall's results [5] [6] [7] are as follows:

$w_M^*$ satisfies the triangular inequality if and only if one of the following conditions is met:

1) $W(M)=1$; 2) $W(M)=2$;

3) $W(M)=3$, and the GNAF of M has one of the following forms ($b>0, c>0$):

$ar^n+\frac{1}{2}(r\pm1)r^{n-1}\mp cr^{n-2}$, $a\geq3$;
$ar^n+(\frac{1}{2}r)r^{n-1}\pm cr^i$, $i\leq n-2$, $a\geq3$, $c\leq\frac{1}{2}r$;
$ar^n+br^{n-1}\pm(\frac{1}{2}r)r^{n-2}$, $a\geq3$;
$2r^n+br^{n-1}\pm cr^i$, $i\leq n-3$, $b\leq\frac{1}{2}r$, $c\leq\frac{1}{2}r$;
$2r^n+br^{n-1}+cr^{n-2}$, $b\leq\frac{1}{2}r$;
$2r^n+br^{n-1}-cr^{n-2}$, $b\leq\frac{1}{2}(r+1)$;
$2r^n+br^{n-1}-(\frac{1}{2}r)r^{n-2}$;
$r^n+br^{n-1}\pm cr^i$, $i\leq n-2$;
$r^n+\frac{1}{2}(r\pm1)r^{n-2}\mp cr^{n-3}$;
$r^n+(\frac{1}{2}r)r^{n-2}\pm cr^i$, $i\leq n-3$, $c\leq\frac{1}{2}r$;
$r^n+br^{n-2}\pm(\frac{1}{2}r)r^{n-3}$;

$ar^n-\frac{1}{2}(r\pm1)r^{n-1}\pm cr^{n-2}$, $a\geq3$;
$ar^n-(\frac{1}{2}r)r^{n-1}\pm cr^i$, $i\leq n-2$, $a\geq3$, $c\leq\frac{1}{2}r$;
$ar^n-br^{n-1}\pm(\frac{1}{2}r)r^{n-2}$, $a\geq3$;
$ar^n-br^{n-2}-(\frac{1}{2}r)r^{n-3}$, $a\geq3$;
$2r^n-br^j\pm cr^i$, $n-j=1$ or $2$, $j-i\geq1$;
$r^n-\frac{1}{2}(r\pm1)r^{n-2}\pm cr^{n-3}$;
$r^n-(\frac{1}{2}r)r^{n-2}\pm cr^i$, $i\leq n-3$, $c\leq\frac{1}{2}r$;
$r^n-br^{n-2}\pm(\frac{1}{2}r)r^{n-3}$;
$3^n-2.3^{n-2}\pm3^i$, $i\leq n-4$;
$r^n-br^{n-3}-(\frac{1}{2}r)r^{n-4}$;
$2^n-2^j\pm2^i$, $3\leq n-j\leq5$, $j-i\geq2$;

4) $W(M)=4$, and the GNAF of M has one of the following forms ($c>0, d>0$):

$r^n+r^{n-1}+\frac{1}{2}(r\pm 1)r^{n-2}\mp dr^{n-3}$;

$r^n+r^{n-1}+(\frac{1}{2}r)r^{n-2}\pm dr^i$, $i\leqslant n-3$, $d\leqslant \frac{1}{2}r$;

$r^n+r^{n-1}\pm(cr^i+(\frac{1}{2}r)r^{i-1})$, $i\leqslant n-2$;

$r^n+r^{n-1}+cr^{n-2}-(\frac{1}{2}r)r^{n-3}$;

$r^n+2r^{n-1}\pm(cr^{n-2}+(\frac{1}{2}r)r^{n-3})$;

$4^n+2.4^{n-2}\pm(4^{n-3}+2.4^{n-4})$;

$3^n+3^{n-2}+3^{n-3}+3^{n-4}$;

$3^n\pm(2.3^{n-2}-3^{n-3}-3^{n-4})$;

$a4^n-2.4^{i+2}\pm(4^{i+1}+2.4^i)$,
 $a=1$ and $n-i=4$, or $a=n-i=3$;

$a3^n-3^{i+2}-3^{i+1}-3^i$,
 $a=1$ and $n-i=4$, or $a=2$ and $n-i=3$.

REFERENCES


[1] T.R.N. Rao, O.N. Garcia *Cyclic and Multiresidue Codes for Arithmetic Operations*. IEEE, Trans. on Inf. Theory, V. IT-17, N° 1, pp 85-91, 1971.

[2] J.L. Massey, O.N. Garcia *Error-Correcting Codes in Computer Arithmetic*. Advances in Information Systems Science, Vol. 4 (Ch.5), pp 273-326, Plenum Press, New York-London, 1972.

[3] W.E. Clark, J.J. Liang *Arithmetic Weight for a General Radix Representation of Integers*. IEEE, Trans. on Inf. Theory, V. IT-19, N° 6, pp 823-826, 1973.

[4] W.E. Clark, J.J. Liang *On Modular Weight and Cyclic Nonadjacent Forms for Arithmetic Codes*. IEEE, Trans. on Inf. Theory, V. IT-20, N° 6, pp 767-770, 1974.

[5] S. Ernvall *When does the Modular Distance induce a Metric in the Binary Case?* IEEE, Trans. on Inf. Theory, V. IT-28, N° 4, pp 665-668, 1982.

[6] S. Ernvall *When does the Modular Distance induce a Metric?* Annales Univ. Turku, Ser. A, Math., N° 185, 1983.

[7] S. Ernvall *On the Modular Distance*. IEEE, Trans. on Inf. Theory, V. IT-31, N° 4, pp 521-522, 1985.

[8] A. Lobstein *When are Modular Weights identical?* EUT Report 86-WSK-05, University of Technology of Eindhoven, Netherlands, 1986.

[9] A. Lobstein *A Note on "A Note on Perfect Arithmetic Codes"*. IEEE, Trans. on Inf. Theory, in press.

# OPTIMUM '1' - ENDED BINARY PREFIX CODES

by

Toby Berger and Raymond Yeung

Departement Systemes et Communications
CNRS, UA 820
Ecole Nationale Superieure des Telecommunications (ENST)
46, r. Barrault
75634 Paris Cedex 13, France

on leave from

Department of Electrical Engineering
Cornell University
Ithaca, NY 14853, U.S.A.

*Abstract* - We have investigated the problem of finding a binary prefix code of minimum average code word length for a given finite probability distribution subject to the requirement that each code word must end with a '1'. We give lower and upper bounds to the performance of the optimum code for any information source; the lower bound is tight in some cases. We also describe an algorithm that generates an optimum code for any information source.

## I. INTRODUCTION

Consider the following game:

A ball is put into one of $K$ boxes. Box $b_i$ has known probability $p_i > 0$ of containing the ball. At each step the player specifies a set of boxes, $B = \{b_{i_1}, b_{i_2}, \cdots, b_{i_n}\}$, and receives a "yes" answer if the ball is in one of the boxes specified and a "no" answer if it is not. The game continues until the player specifies the singleton $\{b_i\}$ which contains the ball. The problem is to find a strategy that minimizes the expected number of steps in the game.

*Definition:* A strategy is non-redundant if for each $j = 1, 2, \cdots$ the set $B_j$ of boxes specified at the step $j$ is a proper subset of either $B_i$ or $B_i^c$ for each $i = 1, 2, \cdots, j-1$, where $B^c = \{b_1, \cdots, b_K\} - B$. It is clear that any redundant strategy can be converted to a non-redundant strategy with the same performance.

*Definition:* A feasible code is a binary prefix code for which each code word ends with a '1'.

It is obvious that, under an appropriate convention there is a one-to-one correspondence between a non-redundant strategy and a feasible code. Therefore, finding an optimum strategy for the game is

This work was supported in part by National Science Foundation under Grants ECS-8204886 and ECS 8521218 and in part by Centre National de la Recherche Scientifique, Paris, France.

equivalent to finding a feasible code for the given probabilities $\{p_i\}$ that is optimum in the usual sense of minimum expected code word length.

## II. THE CHARACTERISTIC INEQUALITIES

The following theorem establishes that, in analogy to the Kraft inequality for prefix codes, there are characteristic inequalities for feasible codes.

*Theorem 1:* There exists a feasible code with $r_j$ code words of length $j$, $j = 1, 2, \cdots, L$, where $r_L \neq 0$, if and only if for all $1 \leq i \leq L$,

$$\sum_{j=1}^{i-1} r_j 2^{-j} + r_i 2^{-(i-1)} \leq 1. \tag{*}$$

Furthermore, if (*) is satisfied, then

$$1 - \left[ \sum_{j=1}^{i-1} r_j 2^{-j} + r_i 2^{-(i-1)} \right] = m 2^{-(i-1)} \tag{**}$$

for some nonnegative integer $m$.

*Proof of Theorem 1:* Suppose (*) is satisfied for all $1 \leq i \leq L$. Construct a binary tree of depth $L$ as follows. Since (*) with $i = 1$ simply guarantees that $r_1 \leq 1$, and there is one '1' node at depth 1, we can pick $r_1$ '1' nodes of order 1. Now assume that we have selected all the terminal nodes of order $j$, $j = 1, 2, \cdots, s-1$. Note that in a complete tree for $j < s$ there are $2^{s-j-1}$ '1' nodes at depth $s$ stemming from each node at depth $j$. Since the '1' nodes at depth $s$ stemming from terminal nodes of order less than $s$ are unavailable (i.e., pruned), the number of '1' nodes available at depth $s$ is

$$2^{s-1} - \sum_{j=1}^{s-1} r_j 2^{s-j-1}.$$

But (*) for $i = s$ implies

$$r_s \leq 2^{s-1} \left[ 1 - \sum_{j=1}^{s-1} r_j 2^{-j} \right]$$

$$= 2^{s-1} - \sum_{j=1}^{s-1} r_j 2^{s-j-1},$$

so there are enough available '1' nodes of order $s$. Since this holds for all $s = 1, 2, \cdots, L$, a feasible code can be constructed with any set of code word lengths that satisfy (*).

Conversely, assume a feasible code with $r_j$ code words of length $j$, $j = 1, 2, \cdots, L$, exists. Among the '1' nodes at depth $i$, $1 \leq i \leq L$,

$$\sum_{j=1}^{i-1} r_j 2^{i-j-1}$$

of them stem from previously selected nodes (i.e., code words of length less than $i$). These pruned '1' nodes (at depth $i$), together with the $r_i$ selected '1' nodes at depth $i$, must add up to a number not larger than $2^{i-1}$, the total number of '1' nodes at depth $i$. Thus for all $1 \leq i \leq L$, we have

$$\sum_{j=1}^{i-1} r_j 2^{i-j-1} + r_i \leq 2^{i-1}$$

which implies

$$\sum_{j=1}^{i-1} r_j 2^{-j} + r_i 2^{-(i-1)} \leq 1.$$

To see that (**) is true if (*) is satisfied, divide both sides of (*) by $2^{-(i-1)}$ and rearrange to obtain

$$2^{i-1} - \sum_{j=1}^{i-1} r_j 2^{i-j-1} - r_i \geq 0.$$

The left hand side of the inequality is a nonnegative integer. Denoting it by $m$, we obtain (**).

## *III. BOUNDS ON THE PERFORMANCE OF AN OPTIMUM FEASIBLE CODE*

We now derive lower and upper bounds for the expected length of an optimum feasible code. The lower bound is tight in certain cases.

*Theorem 2 (Lower Bound):* Let the random variable $U$ have probability distribution $\{p_i,\ 1 \leq i \leq K\}$ with $K \geq 2$ and $p_1 \geq p_2 \geq \cdots \geq p_K$. Then the expected length of any feasible code must be greater than or equal to $H(U) + p_K$, where

$$H(U) = -\sum_{k=1}^{K} p_k \log_2 p_k$$

is the entropy of $U$. Moreover this lower bound is tight in certain cases.

*Proof of Theorem 2:* Consider an optimum feasible code for $U$. Suppose the longest codeword is '$X$1', where '$X$' represents a binary string which may be the empty string. By replacing the codeword '$X$1' by '$X$', we produce an ordinary prefix code for $U$. This derived prefix code for $U$ has expected length at least $H(U)$. Since the code is optimum, '$X$1' must occur with probability $p_k$. Hence the expected length of the optimum feasible code must exceed that of the derived prefix code by $p_k$ and therefore must be greater than or equal to $H(U) + p_k$.

To see that this lower bound can be tight, consider the case $p_1 = p_2 = 0.5$. If we assign the codeword '1' to $p_1$, and the codeword '01' to $p_2$, then the expected length of this code is

$$(0.5 \times 1) + (0.5 \times 2) = 1.5.$$

Since

$$H(U) + p_2 = 1 + 0.5 = 1.5,$$

the lower bound is tight.

Q.E.D.

*Theorem 3 (Upper Bound):* There exists a feasible code with expected length less than $H(U) + 1.5$.

*Proof of Theorem 3:* It is known that there exists a prefix code for $U$ with expected length less than $H(U) + 1$. If for this code the total probability of occurrence of those code words which end with a '0' is not greater than 1/2, then appending a '1' to each of these code words produces a feasible code with expected length less than $H(U) + 1.5$. If the total probability of occurrence of those words in the prefix code which end with a '0' is greater than 1/2, simply switch the 0's and 1's in each code word.

Q.E.D.

We stress that the problem under consideration is not trivial in that appending a '1' to each Huffman code word that ends in 0 does not always produce an optimum feasible code. If $\{p_i\} = \{0.3, 0.3, 0.2, 0.2\}$, for example, then the Huffman-derived code shown in Figure 4b has an average code word length of 2.5, which improves to 2.4 after one optimally reassigns the probabilities to its terminal nodes. However, the code shown in Figure 4a has an average word length of only 2.3.

## *IV. REDUCIBILITY OF A CLASS OF CODES*

A set of numbers $< r_i, i = 1, 2, \cdots, L >$ is called a set of lengths. If it satisfies the characteristic inequalities (*), it is called a valid set. Note that a valid set of lengths specifies a class of feasible codes with equal performance as measured by expected code word length, where the higher probabilities are, of course, assigned to shorter code words. We liberally define equality between a class of codes and its set of lengths, e.g., $R = < r_j, j = 1, 2, \cdots, L >$.

*Definition:* The cardinality of a class of codes $R$, denoted by card($R$), is the total number of code words of a code in $R$.

*Definition:* A class of codes $R = < r_j, j = 1, 2, \cdots, L >$ is reducible at depth $i$, $2 \le i \le L$, if it is possible to move a code word from depth $i$ to depth $i - 1$, i.e., if $< r_j' >$ also is a valid set of lengths, where $r_{i-1}' = r_{i-1} + 1$, $r_i' = r_i - 1$ and $r_j' = r_j$ otherwise. If $R$ is not reducible at any depth, we say it is irreducible.

*Definition:* A class of codes is saturated at depth $i$ if for its set of lengths, equality holds in (*).

It is clear from the proof of Theorem 2 that, if a class of codes is saturated at depth $j$, then for any code in the class, each '1' node at depth $j$ is either a code word or is prefixed by a code word.

*Theorem 4:* If a class of codes $R = < r_j, j = 1, 2, \cdots, L >$ is saturated at depth $L - 1$, then it is not reducible either at any depth $s < L - 1$ or at depth $L$.

*Proof of Theorem 4:* Suppose the class of codes is reducible at some depth $s$. Then the resulting reduced lengths also must satisfy (*) in Theorem 1 for $i = L - 1$. If $s < L - 1$, then

$$\sum_{\substack{j=1 \\ j \ne s-1 \\ j \ne s}}^{L-2} r_j 2^{-j} + (r_{s-1} + 1) 2^{-(s-1)} + (r_s - 1) 2^{-s} + r_{L-1} 2^{-(L-2)} \le 1$$

$$\sum_{j=1}^{L-2} r_j 2^{-j} + (2^{-(s-1)} - 2^{-s}) + r_{L-1} 2^{-(L-2)} \le 1$$

$$\sum_{j=1}^{L-2} r_j 2^{-j} + 2^{-s} + r_{L-1} 2^{-(L-2)} \le 1$$

$$\sum_{j=1}^{L-2} r_j 2^{-j} + r_{L-1} 2^{-(L-2)} \le 1 - 2^{-s} < 1$$

which contradicts the assumption that depth $L-1$ is saturated. If $s = L$, then

$$\sum_{j=1}^{L-2} r_j 2^{-j} + (r_{L-1} + 1) 2^{-(L-2)} \le 1$$

$$\sum_{j=1}^{L-2} r_j 2^{-j} + r_{L-1} 2^{-(L-2)} \le 1 - 2^{-(L-2)} < 1$$

which also contradicts the assumption that depth $L-1$ is saturated.

Q.E.D.

*Theorem 5:* If a class of codes $R = < r_j, j = 1, 2, \cdots, L >$ is unsaturated at depth $L-1$, then it is reducible (at depth $L$).

*Proof of Theorem 5:* To show that it is possible to move a code word from depth $L$ to depth $L-1$, we need only check that the resulting set of lengths satisfies (*) in Theorem 1 for $i = L, L-1$. Since the class of codes is unsaturated at depth $L-1$, (**) implies that

$$1 - \sum_{j=1}^{L-2} r_j 2^{-j} - r_{L-1} 2^{-(L-2)} \geq 2^{-(L-2)},$$

or

$$\sum_{j=1}^{L-2} r_j 2^{-j} + (r_{L-1} + 1) 2^{-(L-2)} \leq 1.$$

This is (*) with $i = L-1$ for the resulting set of lengths.

For the original lengths, (*) in Theorem 1 with $i = L$ assures us that

$$\sum_{j=1}^{L-1} r_j 2^{-j} + r_L 2^{-(L-1)} \leq 1.$$

This is equivalent to

$$\sum_{j=1}^{L-2} r_j 2^{-j} + (r_{L-1} + 1) 2^{-(L-1)} + (r_L - 1) 2^{-(L-1)} \leq 1.$$

This is (*) with $i = L$ for the resulting set of lengths. Hence the class of codes is reducible at depth $L$.

Q.E.D.

## V. *AN ALGORITHM TO SEARCH FOR AN OPTIMUM CODE*

*Definition:* Let $R_1 = < r_{1j}, j = 1, 2, \cdots, L_1 >$ and $R_2 = < r_{2j}, j = 1, 2, \cdots, L_2 >$ be two irreducible classes of codes with $\text{card}(R_1) \leq \text{card}(R_2)$. The composition of $R_1$ and $R_2$, denoted by $C(R_1, R_2)$, is defined by

$$C(R_1, R_2) = < r_j, j = 1, 2, \cdots, L >$$

where $L = \max(L_1, L_2) + 1$; if $L_1 = 1$, then

$$r_1 = 1$$

$$r_j = r_{2j-1}, \qquad 2 \leq j \leq L_2 + 1;$$

if $L_1 > 1$, then

$$r = 0$$

$$r_j = \begin{cases} r_{1j-1} + r_{2j-1} & 2 \leq j \leq \min(L_1, L_2) + 1 \\ r_{2j-1} & L_1 + 1 < j \leq \max(L_1, L_2) + 1, L_2 \geq L_1 \\ r_{1j-1} & L_2 + 1 < j \leq \max(L_1, L_2) + 1, L_1 \geq L_2 \end{cases}$$

*Lemma 2:* An irreducible class with two or more code words is the composition of two irreducible classes.

*Proof of Lemma 2:* For any irreducible class $R$ with $\text{card}(R) \geq 2$, choose a representative in the class such that the upper subtree contains at least as many code words as the lower subtree. If the lower subtree is degenerate, take $R_1 = <1>$; otherwise take $R_1$ as the class represented by the lower subtree. Then take $R_2$ as the class represented by the upper subtree. It is then obvious that $R = C(R_1, R_2)$, that $R_1$, $R_2$

must be irreducible, and that

$$\text{card}(R_1), \text{card}(R_2) \leq \text{card}(R) - 1$$

and

$$\text{card}(R_1) + \text{card}(R_2) = \text{card}(R).$$

Q.E.D.

It would be desirable to have a recursive procedure, analogous to that devised by Huffman for ordinary prefix codes, that rapidly generates the optimum feasible code for given $\{p_i\}$. Unfortunately, we have been unable to discover such a structural procedure. One factor that we believe contributes significantly to the difficulty of the task is that, unlike in a Huffman code, knowledge of all the code words of order $1 \leq i \leq L - 1$ does not uniquely determine the code words of order $L$.

Before describing our strategy to search for an optimum code, we state the following simple observation without proof.

*Observation 1:* Every optimum code belongs to an irreducible class.

To find an optimum code for $\{p_i, 1 \leq i \leq K\}$, begin by consulting a complete list of irreducible classes of codes with cardinality $K$. By Observation 1 we know that an optimum code for $\{p_i\}$ must be a representative of one of the classes of codes on the list. To obtain the shortest expected code word length for some irreducible class on the list, simply assign shorter code words to symbols with higher probability of occurrence. It will be shown in the next paragraph that there is a finite number of irreducible classes of each finite cardinality $K$. Therefore, we can compute the shortest expected length of each irreducible class of codes of cardinality $K$ when used to encode $\{p_i, 1 \leq i \leq K\}$, thereby ascertaining the optimum class of codes for that particular $\{p_i\}$.

It remains to obtain a complete list of irreducible classes of codes for each cardinality. We do this recursively. There is only one irreducible class of codes with cardinality one, namely the class <1>; actually there is only one code in this class. Now assume that we have a complete list of irreducible classes of codes for cardinality up to $K - 1$. In order to construct the list for cardinality $K$, we use our knowledge that each irreducible class of codes $R$ of cardinality $K$ is of the form $R = C(R_1, R_2)$ for some $R_1, R_2$ such that

$$\text{card}(R_1), \text{card}(R_2) \leq K - 1$$

and

$$\text{card}(R_1) + \text{card}(R_2) = K.$$

Arbitrarily choose $R_1$ , $R_2$ which satisfy the above requirements, and check whether or not $C(R_1, R_2)$ is saturated at the penultimate depth. If not saturated, from Theorem 5 we know that $C(R_1, R_2)$ is reducible, so we ignore this class. If saturated, we proceed to check whether $C(R_1, R_2)$ is reducible at the penultimate depth, because from Theorem 4 we know that this is the only depth at which reduction may be possible. If it is not reducible at the penultimate depth, add this class to the list of irreducible classes of cardinality $K$. Then repeat the procedure with different eligible $R_1$ and $R_2$ until all possible combinations of $R_1$'s and $R_2$'s are exhausted. Before adding a new class to the list, make sure that it does not already appear. It is now clear that there is a finite number of irreducible classes for any finite cardinality.

## VI. AN IMPROVEMENT OF THE ALGORITHM

The algorithm presented in Section V is robust in the sense that it can handle all probability distributions. However, there is no guarantee that the algorithm is efficient in the sense that the effort required to search for an optimum code is minimized. Indeed, we can reduce the size of the lists of irreducible classes of codes for cardinalities $K \geq 4$ yet still obtain an optimum code for any probability distribution, so that the time needed to search for the required optimum code is reduced. This improvement is based on the following observation.

*Observation 2:* For cardinality four, the two irreducible classes of codes are <1, 1, 1, 1> and <0, 2, 2>. Denote them by $R_a$ and $R_b$, respectively. Let the probability distribution be such that $p_1 \geq p_2 \geq p_3 \geq p_4$. As usual, higher probabilities are assigned to shorter code words. Then the expected length of the code in $R_a$ is given by

$$EL_a = P_1 + 2p_2 + 3p_3 + 4p_4,$$

while the expected length of the code in $R_b$ is given by

$$EL_b = 2p_1 + 2p_2 + 3p_3 + 3p_4.$$

Therefore,

$$EL_a - EL_b = -p_1 + p_4 \leq 0.$$

As a result, the performance of the code in $R_a$ is always as good as, if not better than, that of the code in $R_b$; equality occurs only when $p_i = 1/4$ for $1 \leq i \leq 4$.

Because of Observation 2, we need never consider $R_b$. Moreover, we need never use $R_b$ in the composition of new classes of higher cardinality because it will always be at least as good to use $R_a$ instead. By ignoring $R_b$ when employing the aforementioned procedure for constructing irreducible classes of codes of higher cardinality, we obtain a new, shorter list for each $K \geq 4$. We shall call the classes on this shorter list *potential* classes. Every potential class is irreducible but not conversely.

A significant bonus afforded by this algorithmic modification is that now we need verify only that a newly composed class of codes of cardinality $K > 4$ is saturated at the penultimate depth in order to ensure that the class is irreducible. This is justified by the following theorem.

*Theorem 6:* If $R = C(R_1, R_2)$ where $R_1$ and $R_2$ are potential classes and card$(R) \geq 5$, then $R$ is irreducible if and only if it is saturated at its penultimate depth.

The proof of the theorem is too lengthy to be presented here. The theorem assures us that a newly composed class is irreducible if it is saturated at its penultimate depth. We could use the characteristic inequality directly to check for said saturation, but the following theorem enables us to do this more efficiently.

*Theorem 7:* Let $R_1 = < r_{1j},\ j = 1, 2, \cdots, L_1 >$ and $R_2 = < r_{2j},\ j = 1, 2, \cdots, L_2 >$ be two irreducible classes. Let $R = C(R_1, R_2) = < r_j,\ j = 1, 2, \cdots, L >$. Then $R$ is saturated at the second highest level if and only if one of the following four mutually exclusive cases prevails :

(i) $L_1 = 1$

(ii) $L_1 = L_2$

(iii) $L_1 \neq 1$, $L_2 = L_1 + 1$ and $R_1$ is saturated at depth $L_1$

(iv) $L_2 \neq 1$, $L_1 = L_2 + 1$ and $R_2$ is saturated at depth $L_2$.

The converse of the theorem is readily proved directly, while it is easier to prove the forward part by contradiction. That is, we prove that if none of (i) - (iv) is true, then $R$ is reducible. If none of (i) - (iv) is true, then either one of the following is true:

(v) $L_2 > L_1 + 1$, $L_1 \neq 1$

(vi) $L_1 > L_2 + 1, L_2 \neq 1$

(vii) $L_2 = L_1 + 1, L_1 \neq 1$ and $R_1$ is unsaturated at depth $L_1$

(viii) $L_1 = L_2 + 1, L_2 \neq 1$ and $R_2$ is unsaturated at depth $L_2$

Again, the proof of the theorem is omited here. The significance of Theorem 7 is that, by keeping track of whether the previously generated potential classes are saturated at the greatest depth, we do not have to use the characteristic inequality to check saturation at the penultimate depth of a newly composed class, thereby avoiding many computations.

## *VII. CONCLUSIONS*

We have compiled a complete list of potential classes of codes of cardinality $K \leq 20$. The number of potential classes of each cardinality up to 20 is given in Appendix 1. Observe that, for cardinalities between 10 and 20, the number of potential classes grows by roughly 1/3 as the cardinality is increased by 1. This suggests that the asymptotic growth rate is exponential rather than algebraic. However, for cardinalities of 20 or less, the complexity of finding an optimum code is acceptable.

The analysis in Section VI shows that the original algorithm presented in Section V is not efficient. However, we do not know if the improved version is efficient, since we do not know if there are potential classes which contain no optimum code for any probability distribution. This seems to be an area that is worth investigating.

## *APPENDIX 1*

| Cardinality | Number of potential classes |
|---|---|
| 1 | 1 |
| 2 | 1 |
| 3 | 1 |
| 4 | 1 |
| 5 | 2 |
| 6 | 3 |
| 7 | 4 |
| 8 | 6 |
| 9 | 8 |
| 10 | 12 |
| 11 | 16 |
| 12 | 23 |
| 13 | 30 |
| 14 | 42 |
| 15 | 55 |
| 16 | 75 |
| 17 | 98 |
| 18 | 131 |
| 19 | 170 |
| 20 | 224 |

# OPTIMALITY OF RIGHT LEANING TREES

Herman AKDAG, Bernadette BOUCHON
CNRS - LAFORIA
Université Paris VI - Tour 45
4 place Jussieu - 75252 Paris Cédex 05 - France

## 1. CONSTRUCTION OF BINARY TREES

The problem of the construction of a binary tree satisfying given constraints can be regarded from different points of view. If the terminal nodes of the tree are supposed to be associated with taxa to be recognized and some binary tests are used in every non terminal node, the sequence of results of tests used on a path coming from the root to any given terminal node of the tree provides the identification of a unique taxon. In the case where any pair of edges beginning in a non terminal node are associated with the letters of a binary alphabet, for instance 0 or 1, every terminal node corresponds to the possible symbols to be transmitted by a given source, encoded by means of the sequence of 0 and 1 associated with the edges of the path going from the root to this terminal node. For a tree describing classes in a population, the non terminal nodes are supposed to be bound with questions asked during an inquiry, the edges coming from each of these points correspond to the possible answers to the considered question and each class is characterized through the list of answers to the questions used between the root and a terminal node.

Such processes can be described by means of valuated trees, with a probability assigned to every node, equal to the sum of the probabilities corresponding to the edges descended from this node. Such trees are particular cases of so-called questionnaires [8], which admit more sophisticated applications. The utilizations of binary trees are various and the above-mentioned situations are provided as examples. In most of the cases, the construction of a tree is required in such way that the corresponding process - identification, encoding or classification for instance - is as efficient as possible, for a given probability distribution $P = \{p_1, ..., p_n\}$ such that $p_1 \geq p_2 \geq ... \geq p_n$ associated with the taxa, symbols or classes connected with the terminal nodes. The process must be run quickly and the usual criterion used to evaluate this efficiency is the length of the tree T, that is to say the average rank of a terminal node (number of edges on a path joining the root to this node) :

$$L(T) = \Sigma_{1 \leq i \leq n} p_i r_i, \quad (1)$$

where $r_i$ represents the rank of the terminal node whith probability $p_i$, $1 \leq i \leq n$.

It is clear that the order of the tests or questions used during the process represented by the tree has an effect on its length and the construction of the tree must be carefully carried out from the root to the terminal nodes, in a top-down way. The only method providing the tree with a minimal length (L-optimal), for a given probability distribution, is the Huffman algorithm, which proceeds from the terminal nodes to the root, in a non natural way.

Several other technics have been exhibited to construct binary trees in the converse way, beginning from the root, with a short length. They generally use operations research methods [2], [7], or they look for trees in which each non terminal node corresponds to a splitting of its probability into two almost equal probabilities assigned to its successors [1], [6]. We propose another method based on some properties of right leaning trees, constructing the graph of the tree by means of a recursive determination and assigning the probabilities systematically.

## 2. L-OPTIMAL TREES

Let the probability distribution P be given. Several conditions are necessary for a tree to be L-optimal. The two most important ones can be expressed as follows : (P1) the nodes -terminal or not- ordered with regard to their non-decreasing probabilities must have non-increasing ranks; (P2) the difference between the ranks of two nodes with the same probability cannot exceed 1. The well-known Huffman tree satisfies these properties and is constructed in the following way : (a) arrange the list LP of probabilities in non-decreasing order, (b) choose the two first ones (q' and q") and suppress them from LP, (c) insert their sum q in LP in preserving the order, (d) construct a non terminal node with probability q and two edges descended from this node, with respective probabilities q' and q", (e) iterate from (b) until LP is empty.

This method provides the absolute minimum of the length L, for a given probability distribution P, but it may exist other trees with the same length. We study some of them, easier to obtain in a systematic way than the Huffman tree.

## 3. PROPERTIES OF RIGHT LEANING TREES

A right leaning tree is a strictly binary tree such that its terminal nodes, listed from the left to the right of the tree, have non increasing probabilities and non decreasing ranks. A particular case of right leaning tree has one terminal node at every rank between 1 and n-2, and two at the last rank (n-1), with the probabilities in non increasing order from the left to the right (Fig. 1). Another particular right leaning tree is the balanced one, with terminal nodes of rank k on the left hand side, and eventually k+1 on the right hand side, with $2^k \leq n < 2^{k+1}$ (Fig. 2).

It must be remarked that any right leaning tree defines a lexicographic code when we assign binary symbols to its edges, for instance 0 to the left-hand edge, 1 to the right-hand one, in every non terminal node : the codes associated with the terminal nodes are non decreasing from the left to the right, with regard to their binary value.

Some right leaning trees present an interest because of their length.

Theorem 1 : For any probability distribution, there exists an L-optimal right leaning tree.

Proof : It is equivalent to prove that there exists a unique right leaning tree R(P) with the same length as a given Huffman tree.

Let us call <u>elementary binary tree (e.b.t.)</u> the graph containing two edges descended from a non terminal node. To construct R(P), we consider a given Huffman tree H(P) with a highest rank of terminal nodes equal to m. At any given rank r from 1 to m, we translate all the e.b.t. from the left to the right, as far as possible and we reorder the probabilities of the terminal nodes of rank r+1 (Fig. 3). The obtained tree is right leaning since H(P) satisfies property (P1).

An interesting problem is the construction of R(P) without the help of H(P). We exhibit some properties of L-optimal right leaning trees which could be used during this construction.

<u>Theorem 2</u> : In an L-optimal right leaning tree, a non terminal node of probability p cannot be the predecessor of a terminal node of probability q and a non terminal node, if $q < p/3$.

<u>Proof</u> : Let us suppose that a node with probability p has a successor x with probability $q < p/3$, which is a terminal node, and another successor y with probability $p-q > 2p/3$. Then, if two nodes are descended from y, at least one of them, say z, has a probability not smaller than $(p-q)/2 > p/3 > q$. The two nodes x and z do not satisfy property P1 and, consequently, the considered right leaning tree is not L-optimal.

<u>Theorem 3</u> : In an L-optimal right leaning tree, a terminal node of probability q descending from a node of probability p is a successor of that node if $q \geq p/2$.

<u>Proof</u> : Let us suppose that a node of probability p has two non terminal successors x and y with respective probability p' and p", such that the greatest probability assigned to a terminal node coming after x is q. Then $p' > q$ and $p'' = p-p' < p-q \leq q$ as soon as $q \geq p/2$. The considered right leaning tree cannot be L-optimal since property P1 is not satisfied for y and the node with probability q. Consequently, if $q \geq p/2$, x is a terminal node.

<u>Theorem 4</u> : In an L-optimal right leaning tree, a terminal node of probability q descending from a node of probability p is a successor of that node if $p/3 \leq q \leq p/2$ and $q+q' > 2p/3$, with q' denoting the greatest element of P not greater than q.

<u>Proof</u> : Let us suppose that the situation is the same as in the proof of theorem 3 for the existence of x and y, and the two greatest probabilities assigned to terminal nodes coming after x are q and q', such that $q \geq p/3$ and $q+q' > 2p/3$. Then $p' \geq q+q'$, $p'' = p-p' \leq p-q-q' < p/3 \leq q$. The right leaning tree cannot be L-optimal since condition P1 is not satisfied with regard to y and the node of probability q. Consequently, if the inequalities concerning q, q' and p are satisfied and the tree is L-optimal, x must be a terminal node.

<u>Theorem 5</u> : If the probability distribution P is such that either $p_1 \geq 1/2$, or $1/3 \leq p_1 \leq 1/2$ and $p_1+p_2 > 2/3$, then $p_1$ will be assigned to a node of rank 1.

<u>Proof</u> : If one of these conditions is satisfied, theorems 3 and 4 imply that $p_1$ is assigned to a terminal node of rank 1 in the L-optimal right leaning tree. Then, it will be the same in a Huffman tree since the probabilities of P are assigned to nodes having the same rank in both of these trees.

We are now able to characterize the L-optimal right leaning tree R(P) among those corresponding to the given distribution P, by means of the previous theorems. In order to obtain a recursive definition of R(P), we propose to represent it by the following method.

## 4. ENCODING OF RIGHT LEANING TREES

For a right leaning tree T with terminal nodes associated with the given probability distribution $P = \{p_1, ..., p_n\}$ such that $p_1 \geq p_2 \geq ... \geq p_n$, we denote by $s_i$ the number of terminal nodes of rank i, with $0 \leq i \leq m$, satisfying

$$\Sigma_{1 \leq i \leq m} s_i = n \text{ and } s_0 = 0 \qquad (2)$$

with $\lceil \log_2 n \rceil \leq m \leq n-1$, if we denote by $\lceil . \rceil$ the upper integer part of the considered real number.

It is clear that the right leaning tree T is uniquely determined by the sequence $S(T) = s_1, ..., s_m$ [3] and the probability distribution P. For instance, there exist the following codewords associated with right leaning trees (Fig. 4):

- for n=2 : S(T) = 2,
- for n=3 : S(T) = 1 2,
- for n=4 : $S(T_1)$ = 1 1 2 or $S(T_2)$ = 0 4.

The <u>length of T</u> is immediatly evaluated by means of the following formula :

$$L(T) = \Sigma_{1 \leq j \leq m}\, j\, \Sigma_{s_0+...+s_{j-1}+1 \leq i \leq s_0+...+s_j}\, p_i. \qquad (3)$$

Then, the construction of an L-optimal right leaning tree R(P) corresponding to a given probability distribution P can be achieved by the determination of all the codewords describing right leaning tree with n terminal nodes $S(T_1)$, $S(T_2)$,... , the evaluation of the length of the associated trees $L(T_1)$, $L(T_2)$, ..., the choice of the smallest value $L(T_\beta)$ and the assignment of probabilities $p_1, ..., p_n$ to the terminal nodes, from left to right, of the corresponding tree encoded by $S(T_\beta)$.

Consequently, it is necessary to give a method to determine all the codewords representing right leaning trees, for a given value of n, and two solutions will be provided in the next section. At first, let us give the bounds of coefficients $s_i$, for $1 \leq i \leq m$.

<u>Theorem 6</u> : A right leaning tree T can be represented by $S(T) = s_1, ..., s_m$ if and only if the following inequalities are satisfied :

(i) $s_i \leq 2^i - \Sigma_{1 \leq j \leq i-1} 2^{i-j} s_j - 1$, for any i such that $1 \leq i \leq m-1$

(ii) $s_m = 2^m - \Sigma_{1 \leq j \leq m-1} 2^{m-j} s_j$

(iii) $s_i \geq 0$, for any i such that $1 \leq i < \lceil \log_2 n \rceil - 1$

(iv) $s_i \geq \max\,[0,\ 2^{i+1} - \Sigma_{1 \leq j \leq i-1} 2^{i+1-j} s_j - n + \Sigma_{1 \leq j \leq i-1} s_j]$, for any i such that $\lceil \log_2 n \rceil - 1 \leq i \leq m-1$.

<u>Proof</u> : A tree with maximum rank of a terminal node equal to i ($i \leq m$) would have at most $2^i$ terminal nodes. Every terminal node of rank $j<i$ removes $2^{i-j}$ such terminal nodes and then the possible values of $s_i$ are given by (i) if $i<m$, since at least a node of rank i must not be terminal for the tree to be continued until rank m, or by (ii) if i=m for the last possible rank of terminal nodes.

For all the indices i such that $1 \leq i < \lceil \log_2 n \rceil - 1$, it is always possible to construct a balanced tree as defined previously, with no terminal node at any rank smaller than $\lceil \log_2 n \rceil - 1$, and $s_i$ can take a zero value for all these indices, as indicated in (iii). If i is at least equal to $\lfloor \log_2 n \rfloor - 1$, i cannot always take a zero value, according to the number of terminal nodes with a rank smaller than i. For a given sequence $s_1, \ldots, s_{i-1}$, the lower bound $s_i^-$ of $s_i$ corresponds to a tree in which the maximum number of nodes of rank i are not terminal and the tree ends at most at the next rank. Then, if i is equal to m-1, (2) together with (ii) yield :

$$\Sigma_{1 \le j \le i-1}\, s_j + s_i^- + (\, 2^{i+1} - \Sigma_{1 \le j \le i-1}\, 2^{i+1-j}\, s_j - 2\, s_i^-\,) = n\,;$$

• if i is different from m-1, $s_i^- = 0$, and (iv) follows immediatly.

Conversely, if conditions (i) to (iv) are satisfied by a sequence of coefficients $s_1, \ldots, s_m$, then a right leaning tree can be defined from a complete binary tree $T^0$ with $2^m$ terminal nodes. We select the $s_1$ left-hand nodes of rank 1, we suppress all the edges and nodes descended from them in $T^0$, we associate with them the $s_1$ greatest probabilities of P, in non increasing order from left to right, we obtain a right leaning tree $T^1$. We repeat this construction for any rank until m-1 and, for instance, for the rank i, we select the $s_i$ left-hand nodes of rank i in $T^{i-1}$, we suppress all the edges and nodes descended from them in $T^{i-1}$, we associate with them the probabilities $p_i$, with i from $s_1+\ldots+s_{j-1}+1$ to $s_1+\ldots+s_j$ in non increasing order from left to right, so that we obtain a right leaning tree $T^i$. Relations (i) to (iv) ensure that this construction is possible.

## 5. CONSTRUCTION OF RIGHT LEANING TREES

A recursive construction of right leaning trees gives all the codewords $s_1, \ldots, s_{m'}$ corresponding to n+1 terminal nodes from those denoted by $s_1, \ldots, s_m$, corresponding to n terminal nodes, with values satisfying conditions of theorem 6.

### 5.1. Redundant construction

The first construction we propose is redundant, but ensures that all the trees are obtained for n+1 terminal nodes from those given for n terminal nodes, and allows us to prove that all the values of coefficients $s_i$, $1 \le i \le m-1$, between the lower and upper bounds given in (i), (iii) and (iv), must be taken into account.

Let us suppose that the family $F_n$ of graphs of right leaning trees with n terminal nodes is given. We can construct the graph of all the right leaning trees with n+1 terminal nodes (let $F_{n+1}$ denote their family) in the following way. For every graph of $F_n$, we consider the last element in the list of terminal nodes of any rank r given from the left to the right, when it exists, and we put an e.b.t. in this node, preserving all the other parts of the considered graph. We get one terminal node less in the obtained graph than in the original one at rank r, and two terminal nodes more at rank r+1, obtaining a graph of $F_{n+1}$. This construction can be precisely defined in the following redundant algorithm :

The only element of $F_2$ is known, as indicated previously. Let us suppose that $F_n$ is given. Then, for any element T of this family, let $S(T) = s_1, \ldots, s_m$ be its associated

codeword. For every coefficient $s_i$ different from zero, with $1 \le i \le m$, we obtain an element of $F_{n+1}$ defined by $S(T') = s'_1, \ldots, s'_{m'}$, with $m' = m$ or $m+1$, such that :
$s'_i = s_i - 1$, $s'_{i+1} = s_{i+1} + 2$, and $s'_j = s_j$ for any j different from i and i+1, $1 \le j \le m$, (4)
and we call this operation the <u>i-th step of the construction using T</u>.

It is clear that there does not exist any other possibility of obtaining an element of $F_{n+1}$, but also that some of them can be obtained several times, from different elements of $F_n$ (see Fig. 5).

<u>Theorem 7</u> : For a given probability distribution P with n elements, the redundant algorithm provides the values of all the possible coefficients $s_i$, for $1 \le i \le m$ and any m between $\lfloor \log_2 n \rfloor$ and $2^n$, satisfying the bounds indicated in relations (i) to (iv).

<u>Proof</u> : We check that all the possible values of coefficient $s_1$ between the bounds indicated in theorem 6 correspond to the only tree existing in $F_2$. Let us suppose that all the values of the coefficients $s_i$ between the bounds given in theorem 6, for $1 \le i \le m-1$, correspond to the definition of a right leaning tree with n terminal nodes. Let us prove that the coefficients $s'_i$ corresponding to right leaning trees with n+1 terminal nodes, for $1 \le i \le m'-1$, obtained by the redundant algorithm, take all the values in the intervals given in theorem 6, for n+1 probabilities.

* For any i such that $1 \le i \le m-1$, there exists at least one $j > i$ such that $s_j \neq 0$, since at least $s_m \ge 2$, and we will obtain $s'_i = s_i$ at the j-th step of the algorithm in such a way that the upper bound given by (i) for $s_i$ is preserved for $s'_i$.

**If $m'=m$, then the values of $s'_m$ come from the j-th steps of the algorithm, for $1 \le j \le m-2$, which preserves the value of $s_m$ for $s'_m$, or for $j=m-1$, in such a way that $s'_{m-1} = s_{m-1} - 1$ and $s'_m = s_m + 2$, and the its value becomes :
$2^m - \Sigma_{1 \le j \le m-1} 2^{m-j} s_j + 2 = 2^m - \Sigma_{1 \le j \le m-1} 2^{m-j} s'_j$, satisfying (i).

**If $m'=m+1$, then the value of $s'_{m'}$ comes from the m-th step of the algorithm with $s'_m = s_m - 1$ and $s'_{m'} = 2$, and the upper bound of $s'_m$ is $s_m - 1 = 2^m - \Sigma_{1 \le j \le m-1} 2^{m-j} s_{j-1}$, according to (ii).

* When $1 \le i < \lfloor \log_2 n \rfloor - 1$, the lower bound of $s_i$ is 0, and will be preserved for $s'_i$ and the fact that all the values between lower and upper bounds of $s_i$ correspond to a tree of $F_n$ entail all the values between lower and upper bounds of $s'_i$ correspond to a tree of $F_{n+1}$. It should be the case for $1 \le i < \lfloor \log_2(n+1) \rfloor - 1$ and a difference appears if $\lfloor \log_2(n+1) \rfloor$ is different from $\lfloor \log_2 n \rfloor$, i.e. $n = 2^l$ for some integer l, then the corresponding bound will not be obtained for $i = \lfloor \log_2(n+1) \rfloor - 1$ but should be. This difference will be corrected in the following part of the proof.

* In the case where $\lfloor \log_2 n \rfloor - 1 \le i \le n-1$, the lower bound of $s_i$ is $\max[0, 2^{i+1} - \Sigma_{1 \le j \le i-1} 2^{i+1-j} s_j - n + \Sigma_{1 \le j \le i-1} s_j]$, which can be different from 0 and entail that the i-th step of the algorithm is always possible ($s'_i = s_i - 1$) ; then, it is easy to check that the lower bound of $s'_i$ is the one given by (iv) for $\lfloor \log_2 n \rfloor - 1 \le i \le n-1$, instead of $\lfloor \log_2(n+1) \rfloor - 1 \le i \le n$. There is again a difference if $n = 2^l$, and we must check that the bound

obtained in this case is 0 as expected in the first part of the proof, i.e. max $[0, 2^{i+1} - \Sigma_{1\le j\le i-1} 2^{i+1-j}s'_j) - (n+1) + \Sigma_{1\le j\le i-1} s'_j] = 0$, for $n=2^l$ and $i = |\log_2 n| - 1$. The second term of this expression can be written $\Sigma_{1\le j\le l-2}\, s_j(1-2^{l-j}) -1$, which is not greater than 0, as expected. Once more, the fact that all the values between lower and upper bounds of $s_i$ correspond to a tree of $F_n$ entail all the values between lower and upper bounds of $s'_i$ correspond to a tree of $F_{n+1}$ .

5.2. Non redundant construction

This construction can be simplified in a second algorithm, defined as follows :

- first of all, we consider every graph of $F_n$ with maximal rank m, and we put it at the right-hand terminal node of an e.b.t., obtaining a graph of $F_{n+1}$ with maximal rank m+1.
- then, for every graph of $F_n$, we consider all the ranks r corresponding to at most 2 terminal nodes and we put an e.b.t. at the right-hand node, preserving all the other parts of the considered graph. We get one terminal node less in the obtained graph than in the original one at rank r, and two terminal nodes more at rank r+1, obtaining a graph of $F_{n+1}$ with the same maximal rank m as the original one. (see Fig. 6).

This construction can be precisely defined in the following non redundant algorithm : Let us suppose that $F_n$ is given and consider any element T of this family, associated with the codeword $S(T) = s_1, ..., s_m$. We construct elements of $F_{n+1}$ defined by $S(T') = s'_1, ..., s'_{m'}$, with m' = m or m+1 in the following way :

- step 0 : $s'_1 = 1$, $s'_i = s_{i-1}$ for every i such that $2\le i\le m-1$.
- step 1 : if $s_1=1$,
  then $s'_1=0$, $s'_2=s_2+2$ and $s'_i=s_i$ for every i such that $3\le i\le m$.
- step j, for every j between 3 and $|\log_2(n+1)|$ :
  if $s_i=0$ for every i such that $1\le i\le j-1$ and $1\le s_j\le 2$,
  then $s'_j=s_j-1$, $s'_{j+1}=s_{j+1}+2$, and $s'_i = s_i$ for every i different from j and j+1. (5)

Theorem 8 : For a given probability distribution P with n elements, the non redundant algorithm provides all the corresponding right leaning trees.

Proof : A method similar to the proof of theorem 7 shows that the non redundant algorithm provides all the codewords associated with right leaning trees of $F_{n+1}$, from those corresponding to right leaning trees of $F_n$, with coefficients taking all the values between the bounds indicated in theorem 6.

We remark that the non redundant algorithm takes into account twice each tree of $F_n$ to construct $F_{n+1}$, once in step 0, and once in only one of the other steps j, corresponding to the index of the first non zero coefficient $s_j$ equal to 1 or 2 in its codeword. In the redundant algorithm, every tree of $F_n$ is taken into account as many times as there exists a non zero coefficient in its associated codeword, which is generally much greater than 2.

## 6. OPTIMAL RIGHT LEANING TREES

The determination of an L-optimal right leaning tree can be regarded from different points of view.

(I) First of all, we can consider that the non redundant algorithm provides the families $F_n$ of graphs of right leaning trees with n terminal nodes, for any n between 1 and an arbitrarily great number N.

Any time we need to construct an L-optimal right leaning tree R(P) associated with a given probability distribution P of n elements, we evaluate the lengths of the trees of $F_n$ deduced from P by (3), choose the smallest value, select the corresponding graph of $F_n$ and determine R(P) by assigning the probabilities $p_1, \ldots, p_n$ of P to its terminal nodes from the left to the right.

(II) A faster research of right leaning trees with a short length could be carried out by restricting ourselves to a length close to that of the Huffman tree. Then, we successively evaluate the length of the trees of $F_n$ until we find one not greater than a given threshold $L_{max}$. By this method, we do not evaluate the lengths of all elements of $F_n$ and we do not need any research of minimum. More precisely, we can use the following thresholds, known to be always greater than the length of a Huffman tree :

- For any distribution P : $L_{max} = H + 1 - 2p_n$, where H denotes the Shannon entropy of P [1].
- If $p_1 \geq 1/2$, then we can consider $L_{max} = H + 2 - p_1 + p_1 \log_2 p_1 + (1-p_1)\log_2(1-p_1)$, as indicated in [5].

Further, theorem 3 entails that $s_1 = 1$ and, consequently, we can eliminate the trees of $F_n$ which do not satisfy this condition before evaluating their length.

- If $p_1 < 1/3$, then we can consider $L_{max} = H + p_1 + 0.0860$ [5], and we can restrict ourselves to the trees of $F_n$ such that $s_1 = 0$, as a consequence of theorem 2.
- If $1/3 \leq p_1 \leq 1/2$ and $p_1 + p_2 > 2/3$, the only acceptable trees in $F_n$ are such that $s_1 = 1$, as indicated by theorem 4, and we can consider the following thresholds [4] :

$L_{max} = H + 3 - 5p_1 - p_1 \log_2 p_1 - (1-p_1)\log_2(1-p_1)$ if $p_1 \geq 0.4505$,

$L_{max} = H + 1 + 1/2(1-p_1) - p_1 \log_2 p_1 - (1-p_1)\log_2(1-p_1)$ if $0.4 \leq p_1 \leq 0.4505$,

$L_{max} = H + p_1 + 0.0860$ otherwise.

- A less strict threshold is suitable to any distribution P and provides more solutions in $F_n$, which are not too far from the optimal one : $L_{max} = H + 1$.

(III) Another method consists in the construction of the graphs of $F_n$ capable of being L-optimal, by the non redundant algorithm, any time we need to construct an L-optimal right leaning tree R(P) associated with a given probability distribution P of n elements.

In order to perform this construction, we successively use the non redundant algorithm from $F_1$ to $F_{n-1}$ and then we use properties of the probability distribution P to construct, by means of the same algorithm, the subfamily $F'_n$ of elements of $F_n$ capable of being L-optimal. Then the evaluation of the lengths of all the trees of $F'_n$ thus obtained allows to construct R(P) as in (I), with much less trees to consider, since a preliminary sorting has already been made. We use the following remarks :

- if $p_1<1/3$, then $s_1=0$ because of theorem 2, in an L-optimal right leaning tree. Consequently, to obtain $F'_n$ from $F_{n-1}$, we do not use step 0 of the algorithm, approximately dividing by 2 the number of trees considered in the construction, with regard to the ordinary construction of $F_n$ from $F_{n-1}$ by the non redundant algorithm.

- if $p_1 \geq 1/2$ or if $1/3 \leq p_1 \leq 1/2$ and $p_1+p_2>2/3$, then $s_1=1$ because of theorem 3 or 4, respectively. Consequently, to obtain $F'_n$ from $F_{n-1}$, we only use step 0 of the algorithm, once more approximately dividing by 2 the number of trees considered in the construction.

The utilization of theorems 2 to 4 with regard to probabilities different from the first one would allow to neglect some of the graphs of $F_n$ used in the previous propositions. Property P2 would also help to eliminate some trees which do not satisfy this necessary condition of L-optimality and, thus, cannot be interesting in a study of right leaning trees with minimum length.

## REFERENCES


[1] Akdag, H. : Performances of an algorithm constructing a nearly optimal binary tree, Acta Informatica 20, 1983.

[2] Akdag, H., Bouchon B. : Right leaning trees with a pseudo-huffmanian length, IEEE Int. Symposium. on Inform. Theory, Ann Arbor, 1986.

[3] Darwiche, J. : Construction de questionnaires arborescents quasi-optimaux, Thèse de 3ème Cycle, Université Paris VI, 1985.

[4] Johnsen, O. : On the redundancy of binary Huffman codes, IEEE Transactions on Inform. Theory 26, 2, 1980.

[5] Gallager, R.G. : Variations on a theme by Huffman, IEEE Transactions on Inform. Theory 24, 6, 1978.

[6] Garey, M.R., Graham R.L. : Performance Bounds on the splitting algorithm for binary testing, Acta Informatica 3, 1974.

[7] Payne, R.W., Preece, D.A. : Identification keys and diagnostic tables, a review, J. Roy. Statis. Soc., Ser. A, 1980.

[8] Picard C.F. : Graphs and questionnaires, North Holland, Amsterdam, 1980.

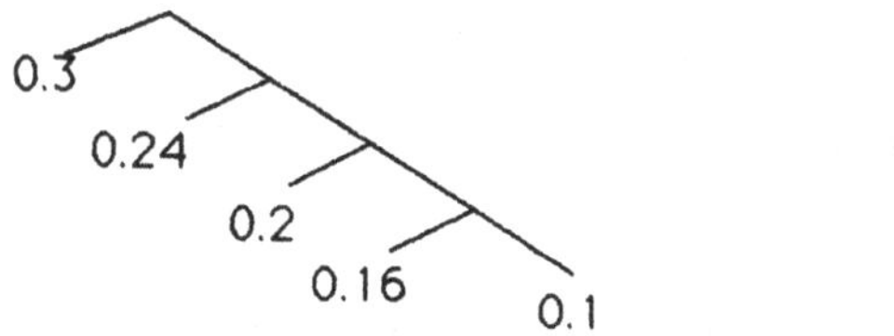


Fig. 1 : Simple binary identification tree

Fig. 2 : Balanced tree

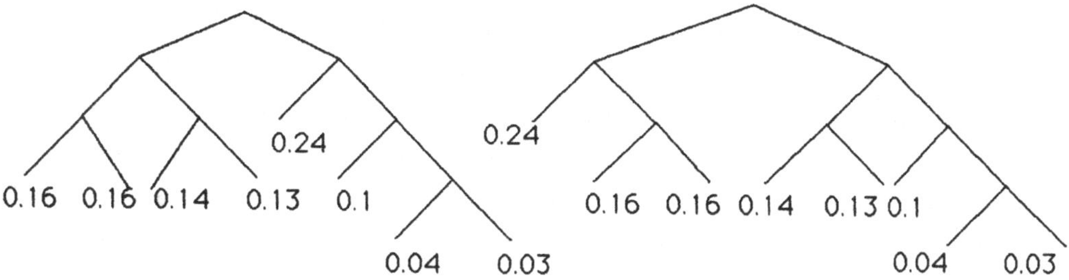


Fig. 3 : Huffman tree and associated right leaning tree with the same length

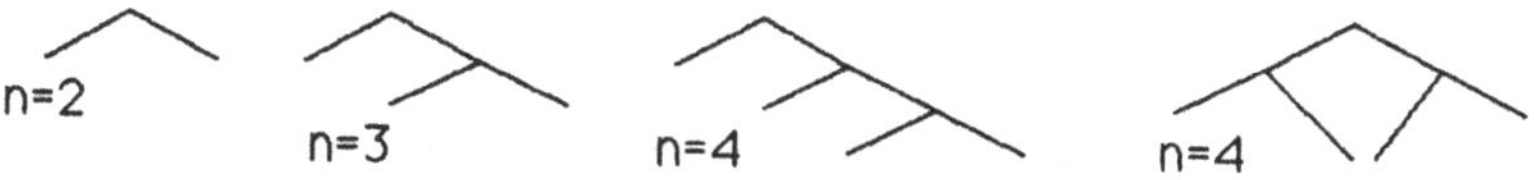


Fig. 4 : Right leaning trees

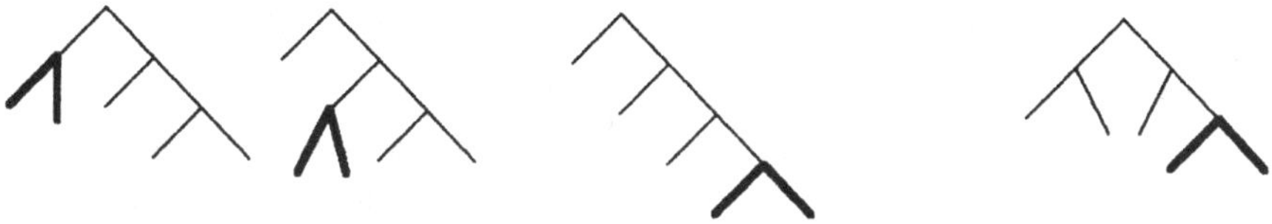

Fig. 5 : Deduction of right leaning trees for n=5 by the redundant algorithm

Fig. 6 : Deduction of right leaning trees for n=5 by the non redundant algorithm

# SPECIFICATION, SYNCHRONISATION, AVERAGE LENGTH.

Anne Bertrand

Université de Bordeaux I

33405 Talence (France)

Abstract. Consider a source providing finite sequences of symbols, zeros and ones in general, into a constrained format dictated by the data processor ; constraints are supposed to be time independant ( of particular importance are those specified by a finite list of forbidden blocks of symbols, e.g. upper and lower bounds on runs length of zeros and ones ) . Factorial extendables languages are an appropriate mathematical model for strings of symbols, and for strings long enough to seem infinite we have to deal with entropy ( Shannon ) if we want to measure the quantities of information dispatched by the source : the greatest the entropy is, the greatest the information will be ; roughly speaking, the entropy is the exponential growth of the number of strings of n letters enclosed in the strings provided by the source.

A particular case is that of a source sending successive blocks $m_1, m_2, \ldots$ randomly chosen in a set X of words on an alphabet A ; we obtain messages of the type $m_1 m_2 \ldots m_n$ ; we call X a variable length code if a word on X has only one decomposition on X ; if $a_n$ is the number of words of n letters in X then the entropy of the associated phenomen is equal to $\log(1/r)$ where r is the positive real number so that : $1 = a_1/r + a_2/r^2 + a_3/r^3 + \ldots$; if you pick up a block m in the sequence dispatched by the source, the average of its length is (Berstel-Perrin) $a = a_1/r + 2a_2/r^2 + 3a_3/r^3 + \ldots$ who is called the average length of the code ; for a given entropy, the smaller the average length is, the more economic the transmission will be and so we investigate the problem : how to minimize the average length of a code for a given entropy ?

Résumé. Considérons une source émettant des suites finies de symboles (des zéros et des uns en général) soumises à certaines contraintes que nous supposerons indépendantes du temps - un cas particulier important consiste à limiter le nombre de répétitions d'un même symbole. Les langages factoriels et prolongeables forment un cadre approprié à l étude de ces suites de symboles ; si les suites sont suffisamment longues nous les considèrerons comme infinies et nous utiliserons la notion d'entropie (Shannon) pour mesurer la quantité d information qu'elles peuvent véhiculer.

Un cas particulier est celui d'une source envoyant des blocs successifs $m_1$ $m_2$, $m_3$ ... choisis aléatoirement parmi un ensemble de mots X sur un alphabet A ; nous obtenons des messages du type $m_1 m_2 \ldots m_k$ ; nous disons que X est un code de longueur variable si un mot sur X se décompose d'une seule façon en produit de mots de X ; si $a_n$ désigne le nombre de mots de longueur n d'un tel code , l'entropie du phénomène associé est égale à $\log(1/r)$ où r est la racine positive de l'équation $1 = a_1/r + a_2/r^2 + a_3/r^3 + \ldots$ ; la longueur d'un mot $m_i$ pris au hasard dans la suite $m_1 m_2 \ldots$ est en moyenne (Berstel Perrin) $l(X) = a_1/r + 2a_2/r^2 + 3a_3/r^3 + \ldots$ ; $l(X)$ est appelée longueur moyenne de X ; pour une entropie donnée , plus la longueur moyenne d'un code est faible, plus la transmission est économique et nous avons donc été amenée à nous intéresser au problème suivant : comment minimiser la longueur moyenne d'un code pour une entropie donnée ?

<u>I-Some vocabulary</u>. Let A be an alphabet, and A the set of words on A ; we say that a word u is a factor of a word v if they are words s and t in A such that v = sut and that u is a left factor of v if v can be written v = us with s in A . We call language every subset of A ; we say that a language L is <u>factorial</u> if every factor of a word of L is still in L, that it is <u>transitive</u> if for every two words u and v in L there is a word s on A with usv in L and that L is <u>extendable</u> if for every word u in L they are letters a and b in A with aub still in A ; all languages we shall consider are supposed to be factorial transitive and extendable (F. T. E. ).

<u>Example</u> : A = {0 , 1 , 2} , L = {0, 1, 22, 221...}; L is not factorial (2 is factor of 22 but is not in L) nor transitive (you cannot go from 0 to 22) nor extendable (how to extend 0 ? on the other hand the language F(L) formed by all factors of words of L are always factorial ; here the words of F(L) are 0 1 2 22 21, 221.

A is always a F.T.P. language.

Take A = 0, 1 and call L' the set of all words on A in witch 1 is always followed by 0 ; this language is clearly F.T.P. Remark that if a word w in L begins and end by zero then you can write w uniqly in the form $w = x_1 x_2 \ldots x_k$ where $x_i$ is 0 or 10 ; if not, w is factor of 0w0 that you can write in this way ; this is a particular case of a more general situation : let X be a <u>code of variable length</u>, that is a part of A such that if $x_1, x_2, \ldots x_k$ and $y_1, y_2, \ldots y_h$ are in X with $x_1 \ldots x_k = y_1 \ldots y_h$ then $k = h$ and for all i $x_i = y_i$ : X' = {0, 10} is a code and so is X'' = {0, 1, 20, 210, 2110 . . . 211...10} ;on the other hand {a ba, aba} is not a code because aba = a ba .

Let X be the set $\{x_1 x_2 \ldots x_h ; x_i \in X\}$ and let L(X) be the set of words factors of X ;

L(X) is a F.T.E. language and we call coded languages this typa of languages. So L = L(0,10) and L(X") is the set of words on $\{0,1,2\}$ in wich 22, 212, 2112, 2111...12,... never occur.

Let Y be the code formed by the words ab, aabb, aaabbb and so on ; L(Y) is the language you use to open and close parenthesis. Square-free words (i. e. words on A with no factor on the form uu, $u \in A$, form a factorial language who is never coded: if it was coded by a code X xx would be square-free for all x of X!

A Fisher-Automaton (Q, f, q) on an alphabet A is a triplet (Q, f, q) where Q is a finite or numerable set called state set, q is an element of Q called garbage, f is an application of Q×A into Q such that for every a in A, $f(\bar{q}, a) = \bar{q}$ ; the automaton reads a word $a_1 a_2 \ldots a_k$ in this way : choose q in Q (we say that the automaton is in the state q) ; if $f(q, a_1) = q_1$ then the automaton goes to the state $q_1$ ; after that he reads the second letter $a_2$ and goes to the state $q_2$ such that $q_2 = f(q_1, a_2)$ and so on up to $a_k$ ; remark that if the automaton is in the state $\bar{q}$ he does not change any-more ; we say that the automaton recognize the word $u = a_1 \ldots a_k$ if they are two states q and q' different of the garbage state such that u take the automaton from the state q to the state q' ; the set of words recognized by the automaton is a language who is always factorial.

We say that an automaton is irreductible if for every two states q and q' different of the garbage there is a word of A who takes the automaton from state q into state q' ; then the language recognized by the automaton is always F.T.E. ; Hansel and Blanchard have proved that a language is coded if and only if there is an irreducible Fisher-automaton wich recognizes all words of this language and only them.

Examples : this automaton recognizes the language L' of words on $\{0,1\}$ without 11 :

$Q = \{q, q', \bar{q}\}$ and $A = \{0,1\}$

$f(q,0) = f(q',0) = q \qquad f(q,1) = q'$

$f(q',1) = f(\bar{q},0) = f(\bar{q},1) = \bar{q}$

the word 11 is going to the garbage q ; if the automaton is in the state q and reads the word 1001010 the automaton goes successively to the states q', q, q, q', q, q' and q where it stops : it is a good word ; note that the automaton goes immediately to the garbage if, starting from the state q', he reads the same word.

This other automaton recognizes the language L" :

$Q = \{q, q', q\}$ and $A = \{0,1,2\}$

$f(q,0) = f(q,1) = f(q',0) = q \qquad f(q,2) = f(q'\ 1) = q'$

$f(q',2) = f(q\ 2) = f(q\ 0) = f(q,1) = q$

The words 22, 212, 2112 ..., 211...12,... are going to the garbage.

2- Synchronisation. We say that a word u of a language L is synchronizing for L if for every pair (v, w) of words on A such vu and wu are both in L and for every word s on A :

$$vus \in L \Longrightarrow wus \in L.$$

Call "future" of a word r the set of all words t such that rt is in L : u is synchronizing for L if as soon as vu is in L, vu has the same future that u ; we say improperly that the future of u does not depend of his past ; we say that a F.T.E. language is synchronizing if it contains a word synchronizing for itself.

Examples : all the words of the language L' are synchronizing for L' : every word of L' ending with 1 can be followed by all word of L' beginning by 0 , and the words finishing with 0 can be followed by all words of L'.

The word 0 is synchronizing for L", but not 1 : after 01 you can put 2 and obtain 012 who is in L" but after 21 you cannot put 2 because 212 is not in L" .

The word ba is synchronizing for the language L(ab aabb aaabbb,...) .

Hansel and Blanchard [1] have proved that a synchronizing F.T.E. language L is always a coded language ;especially if u is a synchronizing word the set of words v such that uv is in L and such that for every two words s and t :

$$sut \in L \Longleftrightarrow svt \in L$$

whose no left factor has the same properties is a code X which L is the associated language L(X) ; we say that X is a synchronizing code for the word u.

Example : For L' the code associated to the word 0 is $\{0, 10\}$ ; for L", the word 2 is synchronizing and the code X is the set of words 0s2 where s is any word on the alphabet $\{0, 1\}$ ; for the language L(ab , aabb, ... ) and the synchronising word ba the code is $\{ba, abba, aabbba, \dots, a^{n-1}b^{n}a \dots\}$.

Synchronizing languages are interesting ones to many points of view : they are recognized by automaton since they are coded languages and we can chose an automaton of degree one, i.e. there is a word u in L and a state q in Q such that if the automaton reads a word finishing by u it goes always into the state q or $\bar{q}$ ; this word u is of course synchronizing ; the arrival of u "put back the meters to zero" and a possible error of lecture or of coding does not pass the word u ; this ensures a certain security of transmission and decoding of messages ; besides if X is a synchronising code relatively to the word u, everytime you see the word u in a string, you know that after u a new word of X is beginning ; so you can hope to reconstitute messages even if you have lost a piece of them.

3- Specification. We say that a F.T.E. language L has the ptoperty of specification if there is an integer k such that for each pair s,t of words of L we can find a word $a = a_1 a_2 \ldots a_h$ with at most k letters so that sat is still in L. The languages L' and L'' have the specification property with k = 1 : if s and t are in L' or L'' so is s0t ; the language L(ab, aabb, aaabbb...) does not have the property of specification because for going from $a^n b$ to a we need at least n-1 consecutive b and n is arbitrarily great. The rational F.T.E. languages (i.e. whose which are recognized by a automaton with a finite state space Q) have the property of specification ; as a matter of fact L and L'' are rational languages.

Languages with the specification property have special probabilistic properties ; for example it is not very likely that words $a_1 \ldots a_m$ of same length appear with very dissimilar frequences in the strings provided by such languages ; in the last paragraph we shall prove that :

Theorem I- If a F.T.E. language has the property of specification, then it is a synchronizing language.

If you consider a synchronizing F.T.E. language together with a synchronizing code defined as in paragraph 2 with a synchronizing word u, you can decide if it has or not the specification :

TheoremII - Let L be an F.T.E. synchronizing language ; then the following conditions are equivalent :

1- L has the property of specification.

2- There is an integer h so that every word $a = a_1 \ldots a_d$ containing d letters is factor of at least one word of X containing at most d+h letters of A.

4- Exemples : g-shift , average length.

Let $(b_1, b_2, \ldots)$ a sequence of natural integers such that for the lexicographical order :

$$\forall n > 1, \quad (b_n, b_{n+1}, \ldots) \leqslant (b_1, b_2, \ldots) \qquad \text{(I)}$$

this implies that no $b_n$ is greater than $b_1$ ; of course this is not sufficient!

Let $L(b_1, b_2 \ldots)$ the language formed by the words $u_1 \ldots u_k$ witch are verifying for the same order :

$$000\ldots0 \leqslant u_1 u_2 \ldots u_k \leqslant b_1 b_2 \ldots b_k$$

$$00\ldots0 \leqslant u_2 \ldots u_k \leqslant b_1 \ldots b_{k-1}$$

$$0 \leqslant u_k \leqslant b_1$$

if the sequence $(b_1 b_2 \ldots b_n \ldots)$ ends with zeros you have to put $<$ in place of $\leqslant$ in the rigth inequalities.

If for exemple the sequence $(b_1, b_2 \ldots)$ is 21111... the language L(2111...) that you obtain by this way is the language L'' because you have to take letters smaller than $b_1 = 2$ and get rid of words 22, 212, 2112 and so on who are greater than 2111... if the sequence is (11000...) the language is L' : 11 cannot appear because 11 < 11 is not true ; if the sequence is (2000...) then the associated language is the set of all words on $\{0, 1\}$.

Let g be the real number greater than 1-there is only one such number-such that

$$1 = \frac{b_1}{g} + \frac{b_2}{g^2} + \frac{b_3}{g^3} + \ldots \qquad \text{(II)}$$

If you take two different sequences $(b_1 b_2 \ldots)$ with condition (I), you have two different numbers g ; conversely for each real number g strictly greater than 1 there is one-and only one-sequence $(b_1 b_2 \ldots)$ verifying (I) and (II) ; this sequence is defined by the recurrence relation :

$$b_1 = [g] \quad \text{and} \quad r_1 = \{g\} \quad ; \text{ for } n \geqslant 1 \quad b_{n+1} = [g r_n] \quad \text{and} \quad r_{n+1} = \{g r_n\}$$

where $[t]$ and $\{t\}$ denote the entire and fractional parts of t.

We will call L(g) the language associated to the sequence $(b_1 b_2 \ldots)$ relative to g ; this language is also defined by the following code :

$$X(g) = \{ b_1 b_2 \ldots b_k a \; ; \; k = 0, 1, 2 \ldots \; ; \; a < b_{k+1} \}$$

$$= \{ 0, 1, \ldots, (b_1 - 1) \; ; \; b_1 0, b_1 1, \ldots, b_1 (b_2 - 1) \; ; \; b_1 b_2 0 \ldots \}$$

For probabilistic and Number-theoretical reasons L(g) is called "language of the g-shift" ; it is rational if and only if the sequence $(b_1 b_2 \ldots)$ is ultimately periodic ; it has the property of specification if and only if the sequence $(b_1 b_2 \ldots)$ end with zeros or if there is an integer p such that for every integer n

$$b_{n+1} b_{n+2} \ldots b_{n+p} \neq 000\ldots0 \qquad b_{n+p+1} \neq 0.$$

The set of such g is of Lebesgue measure zero but nevertheless large enough: its Hausdorf dimension is positive.

L(g) is synchronizing if and only if there is at least one word of the language L(g) wich never appear in the sequence $(b_1 b_2 \ldots)$ ; then it has the specification.

The language L(g) has a quite special property : let $c_n$ be the number of words of length n (i.e. containing n letters) of a code X, and let $d_n$ be the number of words of length n of X ;we call radius of X the radius of convergence of the serie $(d_n z^n)_{n \geqslant 0}$ and we note it $r_{X^*}$;we call average length of X the number (it can be infinite)

$$l(X^*) = \sum_{n \geq 0} n c_n (r_{X^*})^n$$

Recall that $X^*$ is the set of words of the type $x_1 x_2 \ldots x_k$ where all $x_i$ are in X. The radius $r_{X^*}$ is closely related to the quantity of information carried by $X^*$: the entropy of the associated dynamical system (see [5] for the definition) is equal to $\mathrm{Log}(1/r_{X^*})$ ; and when you write an element of $X^*$ who is supposed randomly chosen under the form $x_1 x_2 \ldots x_k$ then the average of the length of the words $x_i$ occuring in this element is precisely $l(X)$ .

The radius $r_{X(g)}$ is equal to $1/g$ and the average length of the code X(g) is

$$l(X(g)) = \frac{b_1}{g} + \frac{2b_2}{g^2} + \frac{3b_3}{g^3} + \ldots$$

With this notations, if $b'_n$ denote the number of words of length n of a code Y :

<u>Theorem III - Let g be a number greater than 1,8 ;let Y be a code of radius $r_{Y^*} = 1/g$ verifying</u>

$$1 = \frac{b'_1}{g} + \frac{b'_2}{g^2} + \frac{b'_3}{g^3} + \ldots$$

<u>(that's a classical equality often verified) ; then</u>

$$l(Y) \geq l(X(g))$$

<u>and the equality is true only if for every n the number $b'_n$ of words of length n of Y is equal to the number $b_n$ of words of length n of X(g).</u>

So X(g) minimize the average length of a code with a given radius(and thus of a given entropy).

If two codes carry the same information, the one with the smaller average length is the more economic ; if g comes near 1 , the information come near 0 and l(g) is growing to infinity ( [6] ) ; if the stated result is true for all g (that is very probable even if we cannot prove it) it confirm the feeling we all have : if you want to speak gibberish, better to employ great words.

If for example $g = 2$, $(b_1 b_2 \ldots) = 2000\ldots$ and $d_n = 2^n$ so $r_{X(2)^*} = 1/2$ and as there is only two words in the code wich length is 1, $l(X(2))$ is equal to $2/2 + 0/4 + \ldots = 1$

If $g = \frac{1+\sqrt{5}}{2}$ then $1 = \frac{1}{g} + \frac{1}{g^2}$ and $(b_1 b_2 \ldots) = (11000\ldots)$ . L(g) is the language L' already seen and

$$l(X(g)) = \frac{1}{g} + \frac{2}{g^2}$$ (this is approximately 1,4).

If $g = \left[\frac{1+\sqrt{5}}{2}\right]^2$ then $(b_1 b_2 \ldots) = 21111\ldots$ ; L(g) is the language L'' and

$$1(g) = \frac{2}{g} + \frac{2}{g^2} + \frac{3}{g^3} + \frac{4}{g^4} \quad \dots$$

Remark the inequalities verified by the sequences $(b_1 b_2 \dots)$ relative to real numbers yields finite words very like to the Lyndon words [2].

5 - Proofs The aim of this paragraph is to prove Theorem I and II and to sketch the proof of Theorem III .

Proof of Theorem I Let L be a language having the property of specification ; let k be an integer such that for every words a and b in L there is a word t of length at most k with atb in L ; fix a and b and call B(a, b) the set of such words t ; of course B(a, b) is a finite set ; if for every $a_1$ and $b_1$ in A with $a_1a$ and $bb_1$ still in L , $B(a_1 a, bb_1)$ is equal to B(a, b), then atb is a synchronising word for L whatever t can be in B(a b) : as a matter of fact if F is the set of the words v with atbv in L for every $a_1$ such that $a_1$atb is in L, we have

$$v \in F \Longrightarrow a_1 atbv \in L \qquad \text{and} \qquad a_1 atbv \in L \Rightarrow atbv \in L \Longrightarrow v \in F$$

so if watb is in L then watbv $\in$ L $\Longrightarrow$ v $\in$ F :atb is synchronising.

If it is not the case they are $a_1$ and $b_1$ with $a_1a$ in L, $bb_1$ in L and $B(a_1 a, bb_1)$ strictly contained in B(a b) ; if $a_1atbb_1$ is not synchronising for each word t of B(a, b ) there is $a_2$ and $b_2$ such that $a_2a_1a$ and $bb_1b_2$are in L and $B(a_2a_1a, bb_1b_2)$ is strictly contained in $B(a_1a, bb_1)$ ; the first set B(a, b) is finite and the sequence of sets is decreasing so this process cannot continue indefinitely : we shall find a synchronising word $a_k \dots a_1 atbb_1 \dots b_k$.

Proof of Theorem II Show that 2) implies 1) . Let a and b in L ; let s and t be two words of X with s = dae and t = fbg where of course e and f are of length at most h ; then st is in X , therefore in L, and as L is factorial aefb is in L wich has the specification with k = 2h .

Show that 1) implies 2) . Let u be a synchronising word X the code associated to u like above ; let a be a word of L ; let s and t be two words of L of length at least k, such that atu and usatu are in L ; from 1 , satu is still in X ; its length is not greater than 2k+length of a : so we can take h = 2k .

Proof of Theorem III in the case og g greater than 2 (in the other cases the proof is similar but a little more complicated). Let $(b_1 b_2 \dots)$ the sequence relative to g, let Y a code of radius 1/g with:

$$1 = \frac{b'_1}{g} + \frac{b'_2}{g^2} + \dots$$

($b'_n$ denote the number of words of length n of Y) ; suppose that the sequences

$(b_1 b_2 \dots)$ and $(b'_1 b'_2 \dots)$ are different ; there is k such that $b_1 \dots b_{k-1}$ is equal to $b'_1 \dots b'_{k-1}$ and $b_k$ greater than $b'_k$ ; if $b_1 \dots b_{k-1} = b'_1 \dots b'_{k-1}$ and $b'_k < b_k$ then : $\frac{b'_1}{g} + \frac{b'_2}{g^2} + \dots + \frac{b'_{k-1}}{g^{k-1}} + \frac{b'_k}{g^k} > \frac{b_1}{g} + \dots + \frac{b_{k-1}}{g^{k-1}} + \frac{1}{g^k} > 1$

because following Parry 6 :

$$1 = \frac{b_1}{g} + \frac{b_2}{g^2} + \dots \qquad \text{with for every h greater than 1 :} \quad \frac{b_{h+1}}{g^{h+1}} + \frac{b_{h+2}}{g^{h+2}} + \dots < \frac{1}{g^h}$$

and so

$$1(Y) - 1(X(g)) = \left(\frac{b'_1}{g} + \frac{2b'_2}{g^2} + \dots + \frac{kb'_k}{g^k} + \dots\right) - \left(\frac{b_1}{g} + \frac{2b_2}{g^2} + \dots + \frac{kb_k}{g^k} + \dots\right)$$

$$= k\frac{b'_k - b_k}{g^k} + \left(\frac{(k+1)\,b'_k}{g^{k+1}} + \frac{(k+2)\,b'_{k+2}}{g^{k+2}} + \dots\right) - \left(\frac{(k+1)b_{k+1}}{g^{k+1}} + \dots\right)$$

$$= \quad -A \qquad\qquad +B \qquad\qquad -C$$

As

$$\frac{b'_{k+1}}{g^{k+1}} + \frac{b_{k+2}}{g^{k+2}} + \dots = \frac{b_k - b'_k}{g^k} + \frac{b_{k+1}}{g^{k+1}} + \frac{b_{k+2}}{g^{k+2}} + \dots \quad \text{because } 1 = \frac{b'_1}{g} + \dots$$

and $1 = \frac{b_1}{g} + \dots$ and $b_1 \dots b_{k-1} = b'_1 \dots b'_k$ , we have

$$B = \frac{(k+1)\,b'_{k+1}}{g^{k+1}} + \frac{(k+2)\,b'_{k+2}}{g^{k+2}} + \dots \geqslant (k+1)\left(\frac{b'_{k+1}}{g^{k+1}} + \frac{b'_{k+2}}{g^{k+2}} + \dots\right)$$

$$\geqslant (k+1)\left(\frac{b_k - b'_k}{g^k} + \frac{b_{k+1}}{g^{k+1}} + \frac{b_{k+2}}{g^{k+2}} + \dots\right)$$

$$B \geqslant A + \frac{1}{g^k} + (k+1)\left(\frac{b_{k+1}}{g^{k+1}} + \frac{b_{k+2}}{g^{k+2}} \dots\right)$$

Let us major the term C :

$$C = \frac{(k+1)b_{k+1}}{g^{k+1}} + \frac{(k+2)\,b_{k+2}}{g^{k+2}} + \dots$$

$$C = (k+1)\left(\frac{b_{k+1}}{g^{k+1}} + \frac{b_{k+2}}{g^{k+2}} + \dots\right) + \frac{b_{k+2}}{g^{k+2}} + \frac{2b_{k+3}}{g^{k+3}} + \frac{3b_{k+4}}{g^{k+4}} + \dots$$

Following Parry (this is related to condition II p. 6) we have for all h :

$$\frac{1}{g^h} > \frac{b_{h+1}}{g^{h+1}} + \frac{b_{h+2}}{g^{h+2}} + \dots$$

and so $\frac{b_{k+2}}{g^{k+2}} + \frac{2b_{k+3}}{g^{k+3}} + \frac{3b_{k+4}}{g^{k+4}} + \dots$ is smallest than $\frac{1}{g^{k+1}} + \frac{1}{g^{k+2}} + \dots = \frac{1}{g^k(g-1)}$

who is equal to $\frac{1}{g^k(g-1)}$. So

$$-A + B - C \geqslant \frac{1}{g^k} - \frac{1}{g^k(k-1)} > 0 \quad \text{if } g > 2$$


[1] Blanchard F. and Hansel G. : Systèmes codés. Th. Comp. Sci. 1986.

[2] Berstel J. and Perrin D. : Theory of codes. Orlando Academic press 1985.

[3] Denker M., Grillenberger C. and Sigmund K. : Ergodic Theory on Compact Spaces ; Lectures Notes in Mathematics , Vol. 527 1976.

[4] Bertrand-Mathis A. Développements en base $\theta$ , répartition modulo un, systèmes codés. Bulletin de la S. M. F. 1968.

[5] Shannon C. A Mathematical Theory of Communication , Bell System Tech. 1946.

[6] Parry W. On the $\beta$-expansions of reals numbers, Acta Math Sci. Hungar. 11 , 1960 .

# AMELIORATION OF THE McWILLIAMS-SLOANE TABLES USING GEOMETRIC CODES FROM CURVES WITH GENUS 1,2 OR 3.

## (following A.M.Barg,G.L.Katsman and M.A.Tsfasman)

by

Yves DRIENCOURT

(Université Aix-Marseille II & C.I.R.M.)

and

Jean Francis MICHON

(Université Paris 7)

### 1 . Introduction

The theory of geometric codes inaugurated by **Goppa** ([3],[4]) has already led to several decisive results in coding theory , whose the best known is the construction of an infinite family of codes with parameters better than the **Varshamov-Gilbert** bound (work of **Tsfasman ,Vladut et Zink** ,[11],[12]) . Beside the study of new families of codes (issued from elliptic curves for example) there exist some less known results showing the power of the geometric tools : among others the amelioration of the table concerning the best known binary codes (of length up to 512 and minimum distance up to 29) appearing in the book by **F.J. McWilliams** and **N.J.A. Sloane** ([8]) and the construction of sphere packings associated with geometric codes .The aim of this paper is to illustrate the first of these topics . To do this , we build a few codes whose parameters are announced in the work of **Barg , Katsman** and **Tsfasman** ([1]).

These codes are obtained in the following way : one considers outer geometric codes over $\mathbf{F}_8$ ou $\mathbf{F}_{16}$ concatenated with inner binary codes as shown by **Zinoviev** ([13]) . The resulting codes are not linear. By the way we must note that this construction had already proved its interest , as is shown in the preceding table where a lot of codes "Zv" can be found.

We present here (see table 1 originated from [1]) three examples where we can recognize as outer geometric codes an elliptic code , a code arising from a curve of genus 2 and finally a code issued from a curve of genus 3 . The different systems of inner codes occuring in the concatenation are also described .

**Table 1**

| d | n | r | $n_{ext}$ | $k_{ext}$ | q | genus | Inner code |
|---|---|---|---|---|---|---|---|
| 25 | 91 | 67 | 23 | 8 | 8 | 3 | (1) |
| 25 | 104 | 72 | 21 | 8 | 16 | 1 | (2) |
| 27 | 159 | 90 | 32 | 17 | 16 | 2 | (3) |

(1) $B^{(0)} = (4,2^3,2)$ (2) $B^{(0)} = (5,2^4,2)$ (3) $B^{(0)} = (5,2^5,1) = \bigcup B_i^{(1)} (5,2^4,2)$

**Table 2 (originated from [8])**

(n = length ; r = length-log(number of codewords) ; d = minimum distance)

| | | |
|---|---|---|
| d=25 | n=87-91 | r=64-68 |
| | n=103-109 | r=74 |
| d=27 | n=159-167 | r=91 |

The amelioration we have in mind (compare with table 2) concerns the redundancy in comparison with two fixed parameters : length and minimum distance .

We propose here to show explicitely how to construct these codes . The effective construction of the outer codes encounters some difficulties : in fact these codes are issued from curves having a lot of rational points over a fixed finite field $\mathbf{F}_q$. We know an upper limit for this number , it is the **Weil** bound

$N \leq q+1+2gq^{1/2}$

and sometimes an exact value for the maximum $N_q(g)$ (for g=1,2 and some other particular cases see the work of **Serre** , [9] , [10]) . But we must know explicitely i.e. by their equation , the curves which are the best from this point of view , and this is not a trivial question ! This problem is not evokated in [1] .

So we begin to discuss this question (§ 2) and give some curves allowing us to build the previous geometric codes . By the way , we recall how to construct these codes (in one case at least this is not very easy because the curve reaching the upper bound appears with non-ordinary singularities) . We recall also the method of **Zinoviev** (§ 3) to obtain concatenated codes , explicitely in the cases which are of interest for our purpose .

## 2 . Stage n°1 : research of the good curves and construction of outer geometric codes

First recall the parameters of a geometric code , which consist in the image in $(\mathbf{F}_q)^n$ of the linear space L(G) (G denote a positive divisor on the curve , defined over $\mathbf{F}_q$ and L(G) the set of all rational functions on the curve with divisor $\geq$ - G)

- $k = deg(G) - g + 1$ si $2g - 2 < deg(G) < n$
- $d \geq n - deg(G)$

The case g=1 concerns the elliptic curves and is well known ([2]) . The maximum number of rational points for such a curve over $\mathbf{F}_{16}$ is 25 and we can take for instance the curve with equation

$$y^2z + yz^2 = x^3 + xz^2 .$$

In the case g>1 ,we must distinguish two classes

- hyperelliptic curves
- non-hyperelliptic curves

(One says that a curve C is **hyperelliptic** if there exist a morphism of degree 2 from C on the projective line $\mathbf{P}^1$) . One can prove that

- all curves of genus 2 are hyperelliptic
- for any $g \geq 3$, there exist both hyperelliptic and non-hyperelliptic curves of genus g .

In particular if g=2 , we get $N_q(2) \leq 2(q+1)$ . **Serre** has shown that

$N_q(2) = 2(q+1)$ if $q= 2,4,8$

$N_{16}(2) = 33.$

**A . A method to get such a curve (g=2)**

For the values 2 , 4 and 8 , we try to obtain a curve of genus 2 with a covering of degree 2 of $\mathbf{P}^l$ . Precisely we write the equation in the following way :

$$y^2 + y = h(x)/k(x) ,$$

where in the right-hand side appears a rational function which , for every $x \in \mathbf{P}^l$, takes values $A \in \mathbf{F}_q$ so that the polynom $y^2 + y = A$ have roots (automatically different) .

**l) q=2. (Serre)**

The only possible value for A is 0 . Because the right-hand side must be zero for $x = 0,1$ et $\infty$, one takes as h(x) the polynom $x^2+x$ and for k(x) a polynom of degree 3 irréducible over $\mathbf{F}_2$ : $x^3+x+1$. The resulting curve has 6 rational points over $\mathbf{F}_2$.

Moreover :

(0,1,0) has multiplicity 3, with irrational tangents $(x^3+xz^2+z^3=0)$ ,

(1,0,0) has multiplicity 2, with tangents $(y+z=0)$ and $(y=0)$ .

This shows on one hand that the genus of the curve is 2 and on the other hand that the smooth curve obtained after the resolution of the singularities will have also 6 rational points over $\mathbf{F}_2$ (the triple point disappears and the double one is replaced by two points).

**2) q=4.**

Possible values for A are 0 et 1. One may consider $h(x) = x^2+x+1$ and $k(x) = x^3+x+1$. The singularities are ordinary , the genus is computed as above (g=2) and the smooth curve obtained has 10 rational points over $\mathbf{F}_4$ (see the rationality of the tangents).

Beyond that , the generalisation is difficult , but in the case of $\mathbf{F}_{16}$ , we know (by other considerations , see **Serre**) a singular curve with 33 rational points :

$$y^2z^3 + yz^4 + x^5 = 0$$

Only one singular point , but not ordinary , occurs (triple tangent z=0) . This is a problem because **Goppa** 's algorithm ([3]) doesn't work for such a curve . In fact , we can avoid this difficulty with an extension of the previous algorithm as it was shown by **D.Lebrigand** ([6]) . To get an outer geometric code [32,17,14] ,one can take for instance $G = 18Q_0$ (where $Q_0 = (0,0,1)$ ). Looking at the form of degree 7 : $y^5z^2$ , whose divisor is

$$(y^5z^2) = 25P_0 + 10P_3$$

($P_3$ denotes the point which is above the singular point (0,1,0) ,obtained after two resolutions of the singularity) , we get as residual divisor

$R = 7P_0 + 10P_3$ .

Then we can follow the standard algorithm of Goppa , with the determination of forms of degree 7 passing through R .

**B . Genus 3**

**Serre** gives the following values :

$N_2(3) = 7$ $\qquad$ $N_4(3) = 14$ $\qquad$ $N_8(3) = 24$ .

Therefore the curves with a maximum number of rational points are the non-hyper-elliptic one (otherwise the number of points would be limited at 6,10 et 18 respectively).

Here we use the fact that a non-hyperelliptic curve of genus $g > 2$ is birationally equivalent to a non-singular curve of degree 2g-2 in $\mathbf{P}^{g-1}$ (called canonical curve). So if g=3, we must look after the non-singular quartics in $\mathbf{P}^2$. In fact there are some which are defined over $\mathbf{F}_2$, for example :

$x^3y + y^3z + z^3x = 0$ ("KLEIN's quartic")

(it is easy to verify that the only solution of this equation which annihilates the three partial derivatives is (0,0,0) ). Therefore we get :

$(x) = P_2 + 3P_3$

$(y) = P_3 + 3P_1$

$(z) = P_1 + 3P_2$

where $P_1 = (1,0,0)$ , $P_2 = (0,1,0)$ et $P_3 = (0,0,1)$.

To obtain an extern geometric code [23,8,13] , we can take for example $G = 10P_3$ and consider the curve $x^4$ whose divisor is

$(x^4) = 4P_2 + 12P_3$ .

Then the residual divisor is given by

$R = 4P_2 + 2P_3$.

and , because $\dim L(G) = 8$ , we have to find 7 curves of degree 4 passing through R . One has only to consider

$x^3z, z^3x, x^2yz, x^2z^2, y^2z^2, xy^2z, xyz^2,$

and verify that the rational functions on the curve obtained when divided by $x^4$ are linearly independant .

## 3 . Stage n°2 : concatenation with binary codes

### A . Recall of the construction ([8],[13])

We use the following codes :

- r codes $A_1, \ldots, A_r$ where $A_i$ is an $(n,N_i,\delta_i)$ code over $\mathbf{F}_{qi}$ , $q_i$ being a power of 2.

- an $(m, q_1q_2 \ldots q_r, d_1)$ binary code $B^{(0)}$ which is the union of $q_1$ disjoint codes $B_{i1}^{(1)}$ $(0 \leq i_1 \leq q_1-1)$ where $B_{i1}^{(1)}$ is an $(m, q_2q_3 \ldots q_r, d_2)$ code with $d_2 > d_1$, himself union of $q_2$ disjoint codes $B_{i1,i2}^{(2)}$ $(0 \leq i_2 \leq q_2-1)$ where $B_{i1,i2}^{(2)}$ is an $(m, q_3 \ldots q_r, d_3)$ code with $d_3 > d_2$ , and so on . Finally , $B_{i1,i2,\ldots,ir-2}^{(r-2)}$ is union of $q_r$ disjoint codes $B_{i1,i2,\ldots,ir-1}^{(r-1)}$ $(0 \leq i_{r-1} \leq q_{r-1}-1)$ each one of them with parameters $(m,q_r,d_r)$

We donote by $b_{i1,i2,\ldots,ir}$ an arbitrary codeword of $B^{(0)}$ . This means that $b_{i1,i2,\ldots,ir}$ belongs to $B_{i1}^{(1)}$ and $B_{i1,i2}^{(2)} \ldots$ , and represents the $i_r^{th}$ codeword of $B_{i1,i2,\ldots,ir-1}^{(r-1)}$.

Therefore we consider the $n \times r$ array where the $j^{th}$ column ${}^t(a_1^{(j)}, \ldots, a_n^{(j)})$ represents an arbitrary vector of the code $A_j$. Recall that $a_i^{(j)} \in \mathbf{F}_{qj}$ and that the elements of this last field may be labeled by the numbers $0 \ldots q_j-1$.

The $i^{th}$ line $L = (a_i^{(1)}, \ldots, a_i^{(r)})$ of the array may be associated with a unique codeword $b_L$ of $B^{(0)}$, i.e a unique binary m-uple . The $n \times m$ binary arrays constructed in this way (considered as binary vectors of length mn) form a new code Z , called a "generalized concatenated code" .

**Theorem** : Z is a $(mn , N_1..N_r , d \geq \min\{d_1\delta_1, . . ,d_r\delta_r\})$ binary code .

**Illustration in the case r=2**

$$\begin{pmatrix} a_1 & a'_1 \\ \cdots & \cdots \\ \cdots & \cdots \\ \cdots & \cdots \\ a_n & a'_n \end{pmatrix} \xrightarrow[0 \leq a'_i \leq q_2-1]{0 \leq a_i \leq q_1-1} \begin{pmatrix} b_{a_1,a'_1} \\ \cdots \\ \cdots \\ \cdots \\ b_{a_n,a'_n} \end{pmatrix} \begin{matrix} \to a'_1\text{th codeword of } B_{a_1}^{(1)} \\ \\ \to a'_n \text{ th codeword of } B_{a_n}^{(1)} \end{matrix}$$

$$\begin{matrix} \downarrow & \downarrow \\ A_1 & A_2 \end{matrix}$$

**B . Concatenation with geometric codes**

**1) Concatenation of the first order**

**Example 1**

extern code : [23,8,13] over $\mathbf{F}_{2^3}$

inner code : $(5,2^3,2)$ over $\mathbf{F}_2$

The extern code having $(2^3)^8=2^{24}$ elements, we get a code $(92,2^{24},26)$ which can be transformed into a code with parameters $(91,2^{24},25)$ by deleting one coordinate . The table of Sloane-McWilliams has been improved because 91-24=67.

**Example 2**

extern code : [21,8,13] over $\mathbf{F}_{2^4}$

inner code : $(5,2^4,2)$ over $\mathbf{F}_2$

the code obtained in the same way has parameters $(104,2^{32},25)$ and the redundancy is now 72 when the best known so far was 74 .

**2) Concatenation of the second order : example 3.**

First extern code $A_1$ : trivial code [32,1,32] over $\mathbf{F}_2$

Second extern code $A_2$ : [32,17,14] over $\mathbf{F}_{2^4}$

First inner code : $(5,2^5,1)$

Second inner codes : two disjoint codes $(5,2^4,2)$ .

Using the same notations as above , we obtain :

| | | | |
|---|---|---|---|
| $n=32$ | $q_1=2$ | $N_1=2$ | $\delta_1=32$ |
| $m=5$ | $q_2=2^4$ | $N_2=2^{4.17}$ | $\delta_2=14$ |

The concatenated code has parameters $(160,2^{69},28)$ . As above this yields a code of length 159 with minimum distance 27 and redundancy 90 , improving the best known result by one .

**Illustration of this concatenation with two stages**

$$\begin{array}{ccc}
\begin{pmatrix} a_1 & a'_1 \\ \cdots & \cdots \\ \cdots & \cdots \\ \cdots & \cdots \\ a_{32} & a'_{32} \end{pmatrix} & \xrightarrow[0 \le a'_i \le 15]{0 \le a_i \le 1} & \begin{pmatrix} b_{a_1,a'_1} \\ \cdots \\ \cdots \\ \cdots \\ b_{a_{32},a'_{32}} \end{pmatrix} \begin{array}{l} \to a'_1\text{th codeword of } B_{a_1}{}^{(1)} \\ \\ \\ \to a'_{32}\text{ th codeword of } B_{a_{32}}{}^{(1)} \end{array} \\
\downarrow \quad \downarrow & & \uparrow \\
A_1 \quad A_2 & & \text{One codeword of the } (160,2^{69},28)
\end{array}$$

# References


[1] BARG A.M., KATSMAN G.L.,TSFASMAN M.A. , **Algebraic geometric codes got from curves of small genui** , preprint.

[2] DRIENCOURT Y., MICHON J.F. , **Elliptic codes over a field of characteristic 2** , à paraître dans le J.of Pure and Applied Algebra.

[3] GOPPA V.D. , **Algebraico-Geometric Codes** , Izv. Akad. Nauk , S.S.S.R. 46 (1982) , = Math. U.S.S.R. Izvestiya 21 (1983) , 75-91.

[4] GOPPA V.D. , **Codes and Information** , Uspekhi Math. Nauk. 39:1 (1984) 77-120 = Russ. Math. Surveys 39:1 (1984) , 87-141.

[5] KATSMAN G.L.,TSFASMAN M.A.,VLADUT S.G., **Modular curves and codes with a polynomial construction** , IEEE Trans. Info. Theory 30 (1984) , 353-355.

[6] LEBRIGAND D. , RISLER J.J. , **Algorithme de Brill-Noether et codage des codes de Goppa** , preprint (Université Paris 6) .

[7] LITSYN S.N., TSFASMAN M.A. , **Algebraic geometric and number theoric packings of spheres in $R^N$** , Uspekhi Math. Nauk. 40 (1985) 185-186.

[8] MAC-WILLIAMS F.J., SLOANE N.J.A. , **The theory of error-correcting codes** , North-Holland , Amsterdam 1977.

[9] SERRE J.P. , **Sur le nombre des points rationnels d'une courbe algébrique sur un corps fini** , C.R. Acad. Sc. Paris 296 (1983) 397-402.

[10] SERRE J.P. , **Nombre de points des courbes algébriques sur $F_q$** , Sém. Th. Nombres Bordeaux (1982-1983) exp.n°22.

[11] TSFASMAN M.A. , **On Goppa codes which are better than the Varshamov -Gilbert bound** , Probl. Peredachi Info., 18 (1982) , 3-6 = Probl. Info. Trans., 18 (1982) ,163-166.

[12] TSFASMAN M.A.,VLADUT S.G.,ZINK T. , **Modular curves, Shimura curves and Goppa codes, better than Varshamov-Gilbert bound** , Math. Nachrichten 109 (1982) , 21-28.

[13] ZINOVIEV V.A. , **Generalized cascade codes** , Probl. Peredachi Info. 12 (1976) 5-15 = Probl. Info. Trans., 12 (1976) 2-9.

A [32,17,14]-GEOMETRIC CODE COMING FROM A SINGULAR CURVE

D. LE BRIGAND : Université P. et M. CURIE ; Equipe Analyse Complexe et Géométrie, UA213 , 45-46, 5ième étage , 4 place Jussieu 75252 PARIS Cedex 05 .

**Abstract** : Using a generalization to singular curves of GOPPA construction for geometric codes [G] , we construct a code with the above parameters .

UN CODE GEOMETRIQUE SUR $\mathbb{F}_{16}$ DE PARAMETRES [32,17,14] CONSTRUIT A PARTIR D'UNE COURBE SINGULIERE

**Résumé** : A partir d'une généralisation à une courbe singulière de la construction des codes géométriques de GOPPA [G] , on construit un code ayant les paramètres cités dans le titre .

**INTRODUCTION**

Using GOPPA construction , algebraic geometry applied to coding theory provides very good codes . In [G] , GOPPA uses plane curves with ordinary singularities.
Here we give an example of a code associated with a plane curve having a non ordinary singular point.

The precise description of the techniques used in this paper can be found in [LB-R] . However we give below some definitions and an elementary approach of the normalization of a singular curve defined on a field K .

## PRELIMINARIES

*ALGEBRAIC CURVE : Let $P^n(K)$ be the projective n-space over K , $(X_0,X_1,\ldots,X_n)$ the projective coordinates in $P^n(K)$ . A projective algebraic curve is an algebraic variety of dimension one in $P^n(K)$ . A rational point of the curve is a point $P=(x_0,\ldots,x_n)$ such that $x_i$ belongs to K for all i .If n is equal to 2 , such a curve is a plane curve ; it has an equation : $F(X,Y,Z)=0$ , where F is a square free polynomial in K[X,Y,Z] , homogeneous in X,Y,Z ( F is a form) . The degree of C is the degree of the form F . We assume that the curve is absolutely irreducible on K , i.e. irreducible on the algebraic closure of K . If n is greater than 2 , a curve C has no longer such a simple equation ; however , in the case we consider below, the curve possesses a local equation $G(u,v)=0$ in a neighbourhood V(P) of a point P of C , if (u,v) are the local coordinates in V(P) at $P=(0,0)$ . Locally C is just like a plane curve .

*SINGULAR POINT : Let C be a plane curve , F the equation of C ; a point P of C is called singular if F and its three derivatives cancel at P . Suppose $P=(0,0,1)$ ; the local equation of C at P is $\underline{F}(x,y)=F(x,y,1)$ ; the multiplicity of P is the degree $r_p$ of the leading form of $\underline{F}$ (homogeneous polynomial of smallest degree of $\underline{F}$ ). If C has $r_p$ distinct tangents at P , the point P is an ordinary singular point.

*GENUS OF A CURVE : The genus g of a plane curve C (and of its normalization if C is singular) is a natural number which "measures" the complexity of C . For instance , if the degree of C is d and if C has no singular points with coordinates in the algebraic closure of K , then $g=(d-1)(d-2)/2$ .

If $K=GF(q)$ , the number $n_0$ of rational points on C is bounded by the WEIL-bound :

$n_0 \leq q+1+2g\sqrt{q}$ .

*RATIONAL DIVISOR : In this paper such a divisor will be a formal finite sum : $D=\sum_i n_i Q_i$ , where $Q_i$ is a rational point on the curve ; if $n_i$ is positive for all i then D is effective ; the support of D is : supp $D=\{Q_i, n_i \neq 0\}$; the degree of D is : deg $D=\sum_i n_i$ . There is an order relation between divisors ; let $D_1=\sum_i n_i Q_i$ , $D_2=\sum_i m_i Q_i$ , we have : $D_1 \geqslant D_2$ if $n_i \geqslant m_i$ for each i .

*NORMALIZATION :Suppose C is a plane curve with a singular point P; blowing up P means replacing C by another curve $C_1$ which is no longer plane but has "better" singularities or is non singular . The curve $C_1$ is birationally equivalent to C . A blowing up is an application $\pi_1: C_1 \_\_ C$ such that $\pi_1$ induces an isomorphism between $C_1-\pi_1^{-1}(P)$ and $C-\{P\}$ . The points Q of $C_1$ such that $\pi_1(Q)=P$ correspond to the tangents to C at P . If $C_1$ is non singular then $C_1$ is the normalization of C ; otherwise we blow a singular point P' of $C_1$ , obtain a curve $C_2$ and so on . The process stops when we have a non singular curve which is the desired normalization $\mathbb{C}$ of C .

*DIVISOR ASSOCIATED TO A FORM : Let h be a form non divisible by the equation of C , H the plane curve defined by the equation h=0 ; let $R_1,....,R_m$ be the common points to F and H ; for each i we look for a system of local coordinates (u,v,w) in a neighbourhood $V(R_i)$ of $R_i$ : in that system we suppose that $R_i=(0,0,1)$ and $\underline{h}(u,v)=0$ is the local equation of H in $V(R_i)$ . We assume also that there is only one point $S_i$ above $R_i$ in the normalization $\mathbb{C}$ of C (it will be so in our example) . A parametrization of C can be found in $V(R_i)$ : $u=p(t)$ , $v=q(t)$ , where p and q are formal series in K[[t]] . Then $ord(S_i,h)$ is equal to the smallest degree of the serie $\underline{h}[p(t),q(t)]$ and the divisor on $\mathbb{C}$ associated with h is : $(h)=\sum_i ord(S_i,h) S_i$ .

*ADJOINT : Let C be a plane curve , $\mathbb{C}$ its normalization , a plane curve H is an adjoint of C if $(h) \geqslant E$ where h=0 is the equation of H (h is a form) and E is a divisor on $\mathbb{C}$ called adjonction divisor (for a definition of E see [F,p190],[LB-R]). Let D be a rational effective divisor on $\mathbb{C}$ , we can define the K-vector space L(D) of the rational functions on C such that $(f) \geqslant -D$ ( if $f = \varphi_1/\varphi_2$ where $\varphi_1$ and $\varphi_2$ are forms of same degree , we have $(f) = (\varphi_1) - (\varphi_2)$ ). The dimension of L(D) is finite and given by the Riemann-Roch theorem . The construction of a basis of L(D) uses the notion of adjoint of the curve C (see [LB-R]) .

**GOPPA CONSTRUCTION**

We recall , briefly , the GOPPA-construction as exposed in [G]. Let K be the finite field GF(q) , where $q = p^r$, p is a prime and C an algebraic plane curve defined and absolutely irreducible on K , of genus g. Let G and D be two rational and effective divisors on the normalization $\mathbb{C}$ of the curve C such that :

a) $D = Q_1 + \dots + Q_n$ ,the $Q_i$ being distinct rational points of $\mathbb{C}$ .

b) deg G < n

c) the supports of D and G are disjoint .

Then the GOPPA code associated with C,D and G is the image under the linear application $\zeta$:

$\zeta : L(G) \text{------>} K^n$

$\zeta(f) = ( f(Q_1), \dots\dots, f(Q_n) )$.

The application $\zeta$ is injective because of the condition c ). The code has the parameters [n,k,d] , where:

* n =deg D

* $k \geqslant \deg G - g + 1$ (with equality if : $\deg G \geq 2g-1$ )

* $d \geqslant n - \deg G$ .

**A [32,17,14]-CODE**

We take : K = GF(16)

$$C : F(X,Y,Z) = Y^2 Z^3 + Y Z^4 + X^5 = 0 .$$

Remark : this curve is quoted in [S]: it has the maximum number of rational points over K within the WEIL-bound . The degree of C is m=5 . The curve posess a singular point , $P_1=(0,1,0)$ , non ordinary , of multiplicity $r_1=3$ on C .

1-NORMALIZATION

The normalization $\mathbb{C}$ of C is obtained after two blowing-ups $\pi_1$ and $\pi_2$ :

* $\pi_1$ - blowing-up $P_1$ on C : in the chart Y=1, let X=X' and Z=Z'X' . Then we have: $\underline{F}(X',X'Z')=X'^3(X'^2+X'Z'^4+Z'^3)=0$ . The transform $C_1$ of C under $\pi_1$ has the local equation: $F_1(X',Z')=X'^2+X'Z'^4+Z'^3=0$ , in the neighbourhood of the point $P_2=(0,0)$ such that $\pi_1(P_2)=P_1$. The point $P_2$ has the multiplicity $r_2=2$ on $C_1$ .

* $\pi_2$ - blowing-up $P_2$ on $C_1$ : let X'=X"Z" and Z'=Z" .Then we have : $F_1(X"Z",Z")=Z"^2(Z"+X"^2X"Z"^3)=0$ . The transform $C_2$ of $C_1$ under $\pi_2$ has the local equation : $F_2(X",Z")=Z"+X"^2+X"Z"^3=0$ , in the neighbourhood of the point $P_3=(0,0)$ . The point $P_3$ is a non singular point of $C_2$ . As the only singular point of $C_1$ was $P_1$ , $C_2$ is non singular and we get then the normalization $\mathbb{C}$ of the curve C .

The genus g of $\mathbb{C}$ (or of C) is equal to : $g = (m-1)(m-2)/2 - \sum_i r_i(r_i-1)/2 = 2$

2-ADJOINT OF C

We can show that the adjoint divisor is equal to $E= 4\,P_3$ ( see [LB-R] for the proof of this assertion ) ; so a plane curve $C_0$ with equation $\varphi_0$ is an adjoint of the curve C if the divisor $(\varphi_0)$ on $\mathbb{C}$ associated to $C_0$ is greater than the adjoint divisor E .

3-RATIONAL POINTS

The curve $\mathbb{C}$ has $n_0$ rational points over K=GF(q) where $n_0$ is bounded by the WEIL-bound : here we obtain : $n_0 \leqslant 33$ . We show now that the WEIL-bound is reached . The curve C possesses 33 rational points over K :

$Q_0=(0,0,1)$ ; $Q_1=(0,1,0)$ $(=P_1)$ ; $Q_2=(0,1,1)$ ;

$Q_3, \ldots\ldots, Q_7$ : $(a,r,1)$ where $a = r, r^4, r^7, r^{10}, r^{13}$ ;

$Q_8, \ldots\ldots, Q_{12}$ : $(a,r^2,1)$ $\quad a = r^2, r^5, r^8, r^{11}, r^{14}$ ;

$Q_{13}, \ldots, Q_{17}$ : $(a,r^4,1)$ $\quad a = r, r^4, r^7, r^{10}, r^{13}$ ;

$Q_{18}, \ldots, Q_{22}$ : $(a,r^5,1)$ $\quad a = 1, r^3, r^6, r^9, r^{12}$ ;

$Q_{23}, \ldots, Q_{27}$ : $(a,r^8,1)$ $\quad a = r^2, r^5, r^8, r^{11}, r^{14}$ ;

$Q_{28}, \ldots, Q_{32}$ : $(a,r^{10},1)$ $\quad a = 1, r^3, r^6, r^9, r^{12}$

( where $r$ is a generator of $F_{16}$) . If we denote by the same letters the non singular rational points of C and the corresponding points of $\mathbf{C}$ , we see that $\mathbf{C}$ has 33 rational points over K : $Q_0$ , $P_3$ , $Q_2$, $Q_3, \ldots\ldots, Q_{32}$ .

4-DIVISORS D AND G

In that paragraph and the following one we use the algorithm exposed in [LB-R] . We consider two rational effective divisors on $\mathbf{C}$ :

* $D = P_3 + Q_2 + Q_3 + \ldots\ldots + Q_{32}$

* $G = 18\, Q_0$ ;

and we want to construct the generator matrix of the code associated with C,D,G . This code is the image under the linear application $\zeta$ :

$$\zeta : L(G) \text{ ------> } K^{32}$$

$$\zeta(f) = (\, f(P_3), f(Q_2), \ldots, f(Q_{32})\,) .$$

Because the divisor G is such that : $\deg G=18 \geq 2g-1=3$ , the parameters of the code are :

$n = 32$

$k = \deg G - g + 1 = 17$

$d \geq n - \deg G = 14$

## 5-BASIS OF L(G)

We consider the K-vector space ■ of the forms of degree 7. A basis of ■ is :

$\Lambda_0 = Z^7$ ; $\Lambda_1 = XZ^6$ ;......................................................................; $\Lambda_7 = X^7$;

$\Lambda_8 = YZ^6$ ; $\Lambda_9 = YXZ^5$;..................................................$\Lambda_{14} = YX^6$;

$\Lambda_{15} = Y^2Z^5$;.........

.................

$\Lambda_{33} = Y^6Z$ ;$\Lambda_{34} = Y^6X$ ;

$\Lambda_{35} = Y^7$ .

The subspace of the forms with equations divisible by F admits the basis :

$\Lambda'_5 = \Lambda_5 + \Lambda_8 + \Lambda_{15}$ ; $\Lambda'_6 = \Lambda_6 + \Lambda_9 + \Lambda_{16}$ ; $\Lambda'_7 = \Lambda_7 + \Lambda_{10} + \Lambda_{17}$ ;

$\Lambda'_{13} = \Lambda_{13} + \Lambda_{15} + \Lambda_{21}$ ; $\Lambda'_{14} = \Lambda_{14} + \Lambda_{16} + \Lambda_{22}$ ; $\Lambda'_{20} = \Lambda_{20} + \Lambda_{21} + \Lambda_{26}$ ;

We factor by this subspace and , in the new subspace , we look for a form $\varphi_0$ such that : $(\varphi_0) \geqslant G + E$ . Let : $\varphi_0 = \Lambda_{30} = Y^5 Z^2$ . The divisor on ■ defined by $\varphi_0$ is :

$$(\varphi_0) = \mathrm{ord}(Q_0, \varphi_0)\, Q_0 + \mathrm{ord}(P_3, \varphi_0)\, P_3$$

A parametrization of C in a neighbourhood of $Q_0$ is :

$x = t$

$y = t^5 + t^{10} + \ldots\ldots$

so : $\mathrm{ord}(Q_0, \varphi_0) = 25$ .

A parametrization of C in a neighbourhood of $P_1$ is :

$x = t^3$

$z = t^5 + t^{10} + \ldots\ldots\ldots$

so : $\mathrm{ord}(P_3, \varphi_0) = 10$

Then : $(\varphi_0) = 25\, Q_0 + 10\, P_3$ and the residual divisor E' is equal to $7\, Q_0 + 10\, P_3$ .

A basis of the subspace of forms $\varphi$ such that $(\varphi) \geqslant E'$ is :

$\varphi_0 = \Lambda_{30} = Y^5 Z^2$ ; $\varphi_1 = \Lambda_{10} = X^3 Y Z^3$ ; $\varphi_2 = \Lambda_{11} = X^4 Y Z$ ;

$\varphi_3 = \Lambda_{12} = Y^2 Z^5$ ; $\varphi_4 = \Lambda_{15} = Y^2 Z^5$ ; $\varphi_5 = \Lambda_{16} = Y^2 X Z^4$ ;

$\varphi_6 = \Lambda_{17} = Y^2 X^2 Z^3$ ; $\varphi_7 = \Lambda_{18} = Y^2 X^3 Z^2$ ; $\varphi_8 = \Lambda_{19} = Y^2 X^4 Z^2$ ;

$\varphi_9 = \Lambda_{21} = Y^3 Z^4$ ; $\varphi_{10} = \Lambda_{22} = Y^3 X Z^3$ ; $\varphi_{11} = \Lambda_{23} = Y^3 X^2 Z^2$ ;

$\varphi_{12} = \Lambda_{24} = Y^3 X^3 Z$ ; $\varphi_{13} = \Lambda_{25} = Y^3 X^4$ ; $\varphi_{14} = \Lambda_{26} = Y^4 Z^3$ ;

$\varphi_{15} = \Lambda_{27} = Y^4 X Z^2$ ; $\varphi_{16} = \Lambda_{28} = Y^4 X^2 Z$ .

A basis of L(G) is :

$\{ \mathbb{1} ; \varphi_1/\varphi_0 ; \ldots\ldots ; \varphi_{16}/\varphi_0 \}$

Let : $\psi_1 = \mathbb{1}$ ; $\psi_1 = \varphi_1/\varphi_0$ ; ......... ; $\psi_{17} = \varphi_{16}/\varphi_0$ .

Then the generator matrix of the code is :

$$\mathbb{M} = ( \psi_i(P_3) , \psi_i(Q_2) , \ldots\ldots , \psi_i(Q_{32}) )_{\, i=1,\ldots,17}$$

When $\mathbb{M}$ is under the canonical form we find a word of weight 14, so the code has the following parameters : [32,17,14] .The matrix $\mathbb{M}$ is equal to($\mathbb{I}_7$ A) where $\mathbb{I}_7$ is the identity matrix and A is :

| | | | | | | | | | | | | | | |
|---|---|---|---|---|---|---|---|---|---|---|---|---|---|---|
| 14 | 4 | 13 | 9 | 14 | 13 | 13 | 2 | 14 | 8 | 13 | 13 | 14 | 4 | 11 |
| 1 | 1 | 7 | 15 | 2 | 11 | 8 | 9 | 3 | 9 | 13 | 10 | 8 | 10 | 14 |
| 8 | 3 | 9 | 14 | 13 | 5 | 15 | 16 | 15 | 1 | 15 | 6 | 13 | 4 | 7 |
| 8 | 10 | 12 | 8 | 7 | 2 | 16 | 9 | 12 | 8 | 2 | 9 | 12 | 5 | 12 |
| 8 | 14 | 2 | 8 | 11 | 15 | 14 | 5 | 5 | 6 | 3 | 8 | 3 | 3 | 8 |
| 8 | 16 | 6 | 16 | 9 | 7 | 7 | 13 | 15 | 3 | 11 | 5 | 12 | 8 | 9 |
| 8 | 8 | 6 | 11 | 15 | 1 | 1 | 3 | 3 | 2 | 16 | 6 | 4 | 9 | 13 |
| 12 | 15 | 5 | 15 | 16 | 4 | 11 | 7 | 15 | 14 | 2 | 7 | 7 | 11 | 5 |
| 12 | 6 | 15 | 1 | 14 | 11 | 9 | 10 | 7 | 5 | 8 | 5 | 15 | 13 | 4 |
| 12 | 9 | 1 | 15 | 12 | 5 | 8 | 3 | 9 | 12 | 12 | 12 | 5 | 5 | 6 |
| 12 | 2 | 16 | 10 | 8 | 13 | 5 | 9 | 5 | 14 | 3 | 12 | 11 | 13 | 3 |
| 12 | 8 | 1 | 9 | 7 | 13 | 4 | 13 | 15 | 6 | 10 | 8 | 13 | 1 | 15 |
| 15 | 9 | 16 | 5 | 6 | 13 | 2 | 2 | 16 | 13 | 7 | 4 | 12 | 8 | 1 |
| 15 | 8 | 1 | 6 | 7 | 2 | 6 | 14 | 1 | 4 | 9 | 12 | 11 | 15 | 9 |
| 15 | 10 | 5 | 4 | 1 | 5 | 9 | 14 | 8 | 16 | 10 | 10 | 2 | 11 | 9 |
| 15 | 15 | 9 | 4 | 3 | 1 | 9 | 2 | 3 | 2 | 3 | 1 | 5 | 8 | 14 |
| 16 | 6 | 12 | 3 | 9 | 11 | 2 | 8 | 14 | 5 | 13 | 4 | 10 | 1 | 7 |

## CONCLUSION

TSFASMAN and al. announce that they use a code with the same parameters as ours (but they don't give the construction ) to make a $[159,2^{69},27]$-code which is "better" than those of the MACWILLIAMS list ( as explain by DRIENCOURT-MICHON : "Codes et courbes" , in the same publication ).

## REFERENCES

[F] W. FULTON .- "Algebraic Curves" ; Lectures Notes , Benjamin , 1969 .

[G] V.D. GOPPA .- " Algebraico-Geometric Codes " ; Math. USSR , Izvestiya , 21 , 1983.

[LB-R] D. LE BRIGAND - J.J. RISLER .- " Algorithme de BRILL-NOETHER et Codes de GOPPA"; to appear in Bull. SMF .

[S] J.P. SERRE .- Résumé du Cours de l'année 1984-1985 ; Annuaire du Collège de France , Paris , 1985 .

# The Generalized Goppa Codes and Related Discrete Designs from Hermitian Surfaces in $PG(3,s^2)$*

I.M. Chakravarti
University of North Carolina,
Chapel Hill

Abstract

A short description is first given of the fascinating use of Hermitian curves and normal rational curves by Goppa in the construction of linear error correcting codes and estimation of their transmission rate (k/n) and error correcting power (d/n) by invoking Riemann-Roch theorem and the subsequent discovery by Tsfasman, Vladut and Zink of a sequence of linear codes in q symbols, which performs better than those predicted by the Gilbert-Varshamov bound for $q \geq 49$.

Next, generalizing Goppa's construction (Goppa 1983, pp. 76-78), several new codes have been constructed by embedding the non-degenerate Hermitian surface $x_0^3 + x_1^3 + x_2^3 + x_3^3 = 0$ of PG(3,4), in a PG(9,4) via monomials and the weight-distributions of these codes have been calculated.

Using the geometry of intersections of a non-degenerate Hermitian surface in $PG(3,s^2)$, by secant and tangent hyperplanes, a family of two-weight projective linear codes have been derived. For s=2, it is shown that the strongly regular graph of this code gives rise to the Hadamard difference sets $v = 2^8$, $k = 2^7-2^3$, $\lambda = 2^6-2^3$ and $v = 2^8$, $k = 2^7 + 2^3$, $\lambda = 2^6 + 2^3$. In fact, the author has now shown that this construction can be extended to derive the Hadamard difference sets $v = 2^{2N+2}$, $k = 2^{2N+1}-2^N$, $\lambda = 2^{2N}-2^N$, $v = 2^{2N+2}$, $k = 2^{2N+1}+2^N$, $\lambda = 2^{2N}+2^N$. This will be reported in another paper.

## Introduction.

A code C is a subset of an n-dimensional vector space $V_n(q)$ over a finite field GF(q). The Hamming weight of a code vector $w(\underline{a})$, $\underline{a} = (a_1,\ldots,a_n)$, $a_i$ in GF(q), is the number of non-zero symbols in $\underline{a}$. The Hamming distance $d_H(\underline{a},\underline{b})$ between two codewords $\underline{a}$ and $\underline{b}$ is the number of positions i, $1 \leq i \leq n$, in which they differ. Let d be the minimum of the distances between pairs of codewords of C, $d = \min_{\underline{a},\underline{b} \in C} d(\underline{a},\underline{b})$ and let $|C| = M$ = number of codewords in C. Then the code C has the parameters (n,M,d). n is called the length of a codeword . If C is a subspace of dimension k, of $V_n(q)$, C is a linear (n,k,d) code, $M = q^k$. For a linear code d is the same as the minimum of the weights of non-zero codewords. A code C with minimum distance d can correct up to [(d-1)/2] errors. Let $G_{k,n} = (g_{ij})$ be a basis of the linear code C(n,k,d). Then G is called a generator matrix of the code. Let $H_{r,n} = (h_{ij})$ be a basis of the null space C the dual linear code $C^{\perp}$ of C, $H\,G^T = 0$. Then H is called a parity check matrix of C.

Ash (1965, p. 130) has argued that if we could synthesize binary linear codes meeting the Gilbert-Varshamov bound, then such codes could be used to maintain any transmission rate up to $1 - H(2\beta, 1-2\beta)$ (which is close to the capacity of a binary symmetric channel for $\beta < \frac{1}{4}$), with an arbitrary small probability of error. Ash's analysis and discussion in the context of binary linear codes can be extended with suitable modification to q-ary linear codes. This explains why coding theorists are so keen on constructing algebraically, sequences of q-ary linear codes which perform better than the Gilbert-Varshamov bound. However, it is known that there does not exist an infinite sequence of primitive BCH codes of length n over GF(q) ($n = q^m-1$) with $\delta$ and R tending to non-zero limits (see, for instance, MacWilliams and Sloane, 1977) ($\delta = \frac{d}{n}$).

## Goppa codes from algebraic curves

Let X be a smooth (non-singular) irreducible projective curve of genus g in the N-dimensional projective space over $\overline{GF}(q)$ (the algebraic closure of GF(q)) and let Q, $P_1,\ldots,P_n$ be n+1 rational points (with coordinates in GF(q)) on X. Let t be an integer such that $2g-2 < t < n$. The linear space L(tQ) with respect to the divisor tQ is the set of rational functions f such that the order f in Q is $\geq -t$. The codewords of the linear code C(n,k,d) over GF(q) are then defined by $(f(P_1),\ldots, f(P_n))$, f in L(tQ). From Riemann-Roch theorem, it follows that $k = t - g + 1$ and $d \geq n-t$. Hence $R = \frac{k}{n} \geq 1 - \gamma = g/n$ and $\delta = \frac{d}{n}$. (See, for instance, Goppa (1984), Vladut, Katsman and Tsfasman (1984.)

Let the functions $f_0 = 1, f_2,\ldots, f_{k-1}$ be a basis of the space L(tQ) and let $g_{ij} = f_i(P_j)$, $i = 0,1,\ldots k-1$, $j = 1, \ldots, n$. Then $G = (g_{ij})$ is a generator matrix of the linear code C (n,k,d). Let $s = t-2g+1$. Then one can show that every set of s columns of G has rank $k-g = (t-g+1)-g = s$. Thus the dual C' of C, has minimum distance $C' \geq s+1$ and has dimension $k' = n-k$. It follows then $R' = \frac{k}{n} \geq 1 - \frac{g}{n} - \frac{d'}{n} = 1-\gamma-\delta'$. The dual codes {C'} can be shown to be the same as the codes {C*} constructed by Goppa (1981, 1982), now known as generalized Goppa codes. The code vectors were defined as vectors of residues of differentials in the linear space $\Omega(\Sigma P_i - tQ)$ which is isomorphic to $L(K + \Sigma P_i - tQ)$, where K is a canonical divisor of degree 2g-2 and dimension g.

The linear codes {C} constructed above are natural generalizations of Reed-Solomon codes which are maximum distance separable (mds) codes or equivalently orthogonal arrays of index unity (see, Bush (1952), Chakravarti (1963), Singleton (1964) and MacWilliams and Sloane (1977), ch. 11).

Higher the ratio of the number of rational points on the curve X to its genus, better is the performance of the code. Vladut and Drinfel'd (1983) have shown that the limit of this ratio A(q) as the genus tends to infinity, does not exceed $\sqrt{q}-1$ and for $q = p^{2\ell}$ there are families of curves over GF(q) with $A(q) = \sqrt{q}-1$ (Ihara (1982), Tsfasman, Vladut and Zink (1982)). For $q = p^2$ then, $\limsup_{n\to\infty} R = \alpha^q(\delta) \geq 1 - (\sqrt{q}-1)^{-1} - \delta$. The asymptotic form of the Gilbert-Varshamov bound states that there exists a sequence of codes such that

$$\lim_{n\to\infty} \sup R = \alpha_q(\delta) \geq 1 - \delta \log_q(q-1) + \delta \log_q \delta + (1-\delta)\log_q(1-\delta).$$

For $\delta_1 < \delta < \delta_2$, where $\delta_1$ and $\delta_2$ are the roots of the equation

$$\delta \log_q(q-1) - \delta \log_q \delta - (1-\delta)\log_q(1-\delta) - \delta = (\sqrt{q}-1)^{-1}),$$

the first lower bound lies above the Gilbert-Varshamov bound. This equation has roots for $q \geq 49$ ($p \geq 7$, $\ell = 1$) (Tsfasman, Vladut and Zink (1982), Tsfasman 1982).

If the sequence of linear codes {C} exceeds the Gilbert-Varshamov bound, the corresponding sequence of dual codes {C'} also does the same.

Goppa (1982) has defined a class of codes called normal codes determined from a pair of divisors $D = \Sigma P_i$ and $G = \Sigma m_Q Q$, where $P_i$ are rational points on a normal curve F and the carriers of the two divisors are disjoint. If Q belongs to some extension field of GF(q), then both $m_Q Q$ and $m_Q \sigma Q$ lie on the divisor G, where $\sigma Q$ is the Frobenius transform of Q. A normal (D,G) code has the parameters $|n - (q+1)| \leq 2g\sqrt{q}$, $r = \deg G - g+1$, $d \geq \deg G - 2g+2$.. The length n of the code does not exceed the number of rational points on F, for which the well known Hasse-Weil estimate is $|n - (q+1)| \leq 2g\sqrt{q}$, where g is the genus of F. In particular, of special interest are curves on which the upperbound $n = q+1 + 2g\sqrt{q}$ is attained.

For every $q = p^{2h}$, p a prime, the curve $x_0^{\sqrt{q}+1} + x_1^{\sqrt{q}+1} + x_2^{\sqrt{q}+1} = o$ over PG(2,q), called a Hermitian curve has genus $g = (q - \sqrt{q})/2$ and n = number of rational points = $q\sqrt{q}+1$. Hence it satisfies the Hasse-Weil bound. The geometries of Hermitian curves in projective planes and Hermitian surfaces in higher dimensional projective spaces have been extensively studied by Bose (1963, 1971), Bose and Chakravarti (1966), Chakravarti (1970, 1971) and Segre (1965, 1967).

Goppa (1981, 1983, 1984) has used the Hermitian curves $x_0^{s+1}+x_1^{s+1}+x_2^{s+1} = o$ over $PG(2,s^2)$ to construct new linear codes. Each one of these codes and their duals are equivalent to certain orthogonal arrays which are extremely useful as designs with a wide range of applications and also as building blocks for other designs such as resolvable and affine resolvable balanced incomplete block designs, and balanced arrays which can be also used as equidistant codes with maximum distance, balanced codes and uniformly packed codes. (see Raghavarao 1971, for definitions of arrays and designs.)

If h is any element of $GF(s^2)$, where s is a prime or a power of a prime, then $\bar{h} = h^s$ is defined to be conjugate to h and h is conjugate to h since $h^{s^2} = h$. A square matrix $H = (h_{ij})$, $h_{ij} = \bar{h}_{ji}$, i, j = 0,1,...,N is called a Hermitian matrix. The set of all points in $PG(N,s^2)$ whose row vectors $\mathbf{x}^T = (x_0,x_1,\ldots,x_N)$ satisfy the equation $\mathbf{x}^T H x^{(s)} = o$, are said to form a Hermitian variety $V_{N-1}$ if H is Hermitian; $\mathbf{x}^{(s)}$ is the column vector whose transpose is $(x_0^s, \ldots, x_N^s)$. The variety $V_{N-1}$ is said to be non-degenerate if H has rank N+1 and its equation can be taken in the canonical form $x_0^{s+1} + \ldots + x_N^{s+1} = o$. If H has rank r+1, then $\mathbf{x}^T H \mathbf{x}^{(s)}$ can be reduced by a non-singular linear transformation, to the canonical form $y_o\bar{y}_o + \ldots + y_r\bar{y}_r$. The number of points in a non-degenerate Hermitian variety $V_{N-1}$ is $(s^{N+1} - (-1)^{N+1})(s^N - (-1)^N)/(s^2-1)$ and the number of points with exactly r non-zero coordinates in $V_{N-1}$ is $\binom{N+1}{r}[(s-1)^{n-1} - (-1)^{n-1}](s+1)^{r-1}/s$. (Bose and Chakravarti, 1966). If N = 2t+1 or 2t+2, then a non-degenerate Hermitian variety $V_{N-1}$ contains flat spaces of dimension t and no higher. The number of u-flats, $o \leq u \leq [(N-1)/2]$ were derived by Chakravarti (1971).

There exists an extensive literature on the three classical geometries - symplectic, orthogonal and unitary geometries andd their associated classical groups (see, for instance, Dembowski, 1968). Geometry of quadric surfaces in projective spaces (orthogonal geometry) have been used by Bose (1961), Robillard (1969), Hill (1978) and Wolfmann (1977) for constructing linear codes. Delsarte and Goethals (1975) have used symplectic geometry (geometry of alternating forms) to construct linear codes. Further connections between Reed-Muller codes and symplectic forms are now well-known (see, for instance, MacWilliams and Sloane, 1977). Goppa (1981) seems to be the first one to have used a Hermitian curve $x_o^3 + x_1^3 + x_2^3 = o$ over PG(2,4) to construct a new linear code.

One of the objectives of this research program is to find an elementary construction of the algebraic curves or surfaces in projective spaces which provide sequences of codes that perform better than the Gilbert-Varshamov bound.

<u>New codes, symmetric designs, Hadamard difference sets from Hermitian varieties</u>.

As we move from algebraic curves in projective planes to algebraic surfaces in projective spaces of higher dimensions, the applicable part of algebraic geometry to the construction of codes and related designs, become rather complex. In order to be able to construct codes and designs and calculate their parameters, based on say, quadric hypersurfaces, Hermitian varieties and symplectic forms, one has to find out properties of such geometrical objects or, if feasible, use a computer. The geometry of quadric hypersurfaces in PG(r,q) has been studied by Primrose, Segre, Ray-Chaudhuri, Barlotti, Tallini, Panella, Hirschfeld (for references see for instance, Barlotti (1965), Dembowski (1968) and Segre (1967)). The geometry of Hermitian varieties in $PG(N,q^2)$ has been studied by Segre (1965, 1967) Bose (1963, 1971), Bose and Chakravarti (1966), Chakravarti (1971) and others (see, for instance, Dembowski, 1968).

By embedding the Hermitian curve $x_0^3 + x_1^3 + x_2^3 = 0$ of PG(2,4), in PG(5,4) via the linear system of the monomials $x_0^2$, $x_1^2$, $x_2^2$, $x_0x_1$, $x_0x_2$, $x_1x_2$, Goppa constructed (Goppa 1983, pp. 77-78) a linear (D,G) code with n=9, k=3, d=6. We extend here, Goppa's construction by embedding the Hermitian surface $x_0^3 + x_1^3 + x_2^3 + x_3^3 = 0$ of PG(3,4) in PG(9,4) via the linear system of 10 monomials $x_ix_j$, i,j=0,1,2,3. We thus derive a linear code with n=45, k=35, d=4 on q=4 symbols.

A parity check matrix of the form 10 x 45, was constructed from a Hermitian surface $x_o^3 + x_1^3 + x_2^3 + x_3^3 = 0$ in PG(3,4). The columns of the matrix were labelled by the 45 points on the surface and the rows were labelled by the 10 monomials $x_o^2$, $x_1^2$, $x_2^2$, $x_3^2$, $x_ox_1$,

$x_0x_2$, $x_0x_3$, $x_1x_2$, $x_1x_3$, $x_2x_3$ and the entries were the values of the monomials at the points. A computer program written by Mr. R. Tobias, a graduate student, generated the matrix and also found that every set of 3 columns were linearly independent but that there were sets of 4 columns which were dependent. The parity check matrix generates then an orthogonal array $(4^{10},45,4,3)$. Its parameters as a linear code $C^{\perp}$ orthogonal to the former C, are n=45, k=10. Using programs for a personal computer, written by Paul P. Spurr (1986) in his Master's project, the weight distribution of this code has been found. It is $A_0=1$ $A_{22}=2160$, $A_{24}=2970$, $A_{26}=4320$, $A_{28}=40,500$, $A_{30}=122,976$, $A_{32}=233,415$, $A_{24}=285,120$, $A_{36}=233,400$, $A_{38}=97,200$, $A_{40}=20,574$, $A_{42}=4320$, $A_{44}=1620$, with all other $A_i$ equal to zero. ($A_i$ is the frequency of codewords of weight i). The code has minimum distance 22 and hence corrects all error patterns of weight 10 or less. It is an even weight code (that is all its codewords have even weights) although it is not a self-orthogonal nor a formally self-orthogonal code.

We have worked out the weight distributions of three other codes: $C_1$ (n=18, k=10, d=3), its orthogonal $C_1^{\perp}$ (n=18, k=8, d=6) and $C_2$(n=27, k=10, d=8) over GF(4). The columns of the 10 x 18 generator matrix of $C_1$ corresponds to the 18 points of the Hermitian surface $V_2$: $x_0^3 + x_1^3 + x_2^3 + x_3^3 = 0$ in PG(3,4), which have at least one coordinate equal to zero and the rows correspond to the 10 monomials of degree 2: $x_0^2$, $x_1^2$, $x_2^2$, $x_3^2$, $x_0x_1$ $x_0x_2$, $x_0x_3$, $x_1x_2$, $x_1x_3$, $x_2x_3$. Its weight distribution is $A_0 = 1$, $A_1=A_2=0$, $A_3=18$, $A_4= A_5=0$, $A_6=540$, $A_7=810$, $A_8=2295$, $A_9=17,238$, $A_{10}=40,581$, $A_{11}=84,078$, $A_{12}=152,658$, $A_{13}=204,660$, $A_{14}=221,022$, $A_{15}=185,382$, $A_{16}=100,278$, $A_{17}=31,590$, $A_{18}=7425$. Thus the minimum distance is 3 and $C_1$ is a single-error correcting code. Its orthogonal $C_1^{\perp}$ (n=18, k=8) has the weight distribution $A_0=1$, $A_1=A_2=A_3=A_4=A_5=0$, $A_6=144$, $A_7=0$, $A_8=459$, $A_9=672$, $A_{10}=2196$, $A_{11}=6264$, $A_{12}=6750$, $A_{13}=11,016$, $A_{14}=17,388$, $A_{15}=12,744$, $A_{16}=5022$, $A_{17}=194,490$, $A_{18}=216$. This code has minimum distance 6 and corrects all error patterns of weight 2 or less.

The columns of the 10 x 27 generator matrix of $C_2$ corresponds to those 27 points of the Hermitian surface $V_2$ in PG(3,4), which have every coordinate non-zero and the 10 rows correspond to the 10 monomials of degree 2 in the coordinate variables $(x_0,x_1,x_2,x_3)$. Its weight distribution is $A_0=1$, $A_1=A_2=A_3=A_4=A_5=A_6=A_7=0$, $A_8=3$, $A_9=45$, $A_{10}=24$, $A_{11}=141$, $A_{12}=717$, $A_{13}=2031$, $A_{14}=5325$, $A_{15}=14,910$, $A_{16}=31,605$, $A_{17}=61,803$, $A_{18}=105,714$, $A_{19}=152,484$, $A_{20}=182,193$, $A_{21}=179,745$, $A_{22}=146,529$, $A_{23}=95,226$, $A_{24}=48,585$, $A_{25}=16,392$, $A_{26}=4104$, $A_{27}=999$. This code has minimum distance 8 and hence corrects all error patterns of weight 3 or less.

A non-degenerate Hermitian variety $V_2$: $x_0^{s+1} + x_1^{s+1} + x_2^{s+1} + x_3^{s+1} = 0$ in $PG(3,s^2)$ has $(s^3+1)(s^2+1)$ points. It is known (Bose and Chakravarti, 1966) that a plane of $PG(3,s^2)$ meets $V_2$ either in a non-degenerate $V_1$ which consists of $(s^3+1)$ points of the plane or in a degenerate $V_1$ of rank 2 which consists of $s^3+s^2+1$ points of the plane. In the former case, the plane may be called a secant plane and in the latter case the plane is called a tangent plane. Thus the code generated by the $4 \times (s^3+1)(s^2+1)$ matrix whose columns correspond to the $(s^3+1)(s^2+1)$ points of $V_2$ and rows to the four coordinates, is a two-weight projective linear code over $GF(s^2)$ with $n = (s^2+1)(s^3+1)$ and $w_1=(s^2+1)(s^3+1) - (s^3+s^2+1) = s^5$ and $w_2 = (s^2+1)(s^3+1) - (s^3+1) = s^5+s^2$ and the frequencies $A_{w_1} = (s^3+1)(s^4-1)$ and $A_{w_2} = (s^4-1)(s^4-s^3)$. The set of points $\bar{V}_2$ complementary to $V_2$ in $PG(3,s^2)$ also gives rise to a two-weight projective linear code over $GF(s^2)$ with $n = s^3(s^3-s^2+s-1)$, $w_1=s^5(s-1)$, $w_2=s^2(s^4-s^3-1)$ $A_{w_1} = (s^3+1)(s^4-1)$ and $A_{w_2}=(s^4-1)(s^4-s^3)$.

Thus for $s=2$, we get two two-weight projective linear codes over $GF(4)$. The parameters of the code $C_3$ corresponding to the 45 points of the Hermitian surface $V_2$ are $n=45$, $k=4$, $d=32$ and the frequency distribution of the weights of this code is $A_0=1$, $A_{32}=135$, $A_{36}=120$ and all other $A_i=0$. The graph on $v=4^4=256$ vertices corresponding to this code is strongly regular and its adjacency matrix $A=B_2-B_1$ has the eigenvalues $\rho_0=-15$, $\rho_1=17$, $\rho_2=-15$. As a two-class association scheme its parameters are $n_1=135$, $n_2=120$, $p^1_{11}=70$, $p^1_{12}=64$, $p^2_{11}=72$, $p^2_{12}=63$, $p^1_{22}=p^2_{22}=56$. This last equality implies that $B_2$ (the association matrix of the second associates) is the incidence matrix of a symmetric BIB design ($v=256$, $k=120$, $\lambda=56$) and $2B_2-J$ is a Hadamard matrix of order $2^8$ which corresponds to the Hadamard difference set $v=2^8$, $k=2^7-2^3$, $\lambda=2^6-2^3$. Note that $I + B_1$ is the incidence matrix of the complementary symmetric BIB design ($v=256$, $k=136$, $\lambda=72$). (See Bose and Mesner 1959, Delsarte 1972, Raghavarao 1971 for definitions of strongly regular graphs association schemes, BIB designs and difference sets).

The code $\bar{C}_3$ corresponding to the 40 points of $\bar{V}_2$ the set complementary to $V_2$ in $PG(3,4)$ has the parameters $n=40$, $k=4$, $d=28$ and the weight-distribution $A_0=1$, $A_{28}=120$, $A_{32}=135$, and all other $A_i=0$. The adjacency matrix $\bar{A}=\bar{B}_2-\bar{B}_1$ of the strongly regular graph on $v=256$ vertices associated with this code has the eigenvalues $\rho_0=15$, $\rho_1=15$, $\rho_2=-17$. $\bar{B}_1$ (the association matrix of the first associates in this 2-class association scheme is the incidence matrix of the symmetric BIB design ($v=256$, $k=120$, $\lambda=56$) and $I + \bar{B}_2$ is the incidence matrix of the SBIB ($v=256$, $k=136$, $\lambda=72$).

Since a projective $(n,k)$ code over $GF(s^r)$ with weights $w_i$ $i=1 \dots t$ determines a projective $(n',k')$ code over $GF(s)$ with weights $w'_i$ $i=1,\dots t$, $n' = n(s^r-1)/(s-1)$, $k'=kr$, $w'_i=s^{r-1}w_i$, $i=1,\dots t$, (Delsarte, 1972), the two projective two-weight codes $C_3$ and $\bar{C}_3$ over $GF(4)$ give rise to two projective binary two-weight codes $C_4$ and $\bar{C}_4$ (say) respectively. The two-weight binary code $C_4$ corresponding to $C_3$, has the parameters $n'=135$, $k'=8$, $w'_1=64$, $w'_2=72$, $A_0=1$, $A_{64}=135$, $A_{72}=120$. Then the strongly regular graph associated with this code

$C_4$ has the same parameters as these of the graph of $C_3$. Consider the 120 code-vectors each of weight 72, which are non-adjacent to (second associates of) the null code-word. Then since $p^1_{22} = p^2_{22} = 56$, it follows that these 120 code-vectors form a difference set, $v = 2^8$, $k = 2^7-2^3$, $\lambda = 2^6-2^3$. On the other hand the 135 code vectors each of weight 64 which are adjacent to (first associate of) the null codeword , together with the null vector gives rise to the difference set ($v = 2^8$, $k = 2^7+2^3$, $\lambda = 2^6+2^3$) since $2+p^1_{11} = p^2_{11} = 72$.

The author has now generalized the above construction to the case of the intersections of a non-degenerate Hermitian variety $V_{N-1}$ by the hyperplanes of $PG(N,2^2)$ for every $N > 1$. This construction provides the sequence of two-weight linear codes over GF(4) with parameters $n = (2^{2N+1} + (-2)^N)/3$, $k = N+1$, $w_1 = 2^{2N-1}$, $w_2 = 2^{2N-1} + (-2)^{N-1}$, $A_{w_1} = n_1 = 2^{2N+1} + (-2)^{N+1} + (-2)^N - 1$, $A_{w_2} = n_2 = 2^{2N+1} + (-2)^N$.

Using the associated strongly regular graph one gets the incidence matrices of the sequence of symmetric BIB designs with $v = 2^{2(N+1)}$, $k = 2^{2N+1}-2^N$, $\lambda = 2^{2N}-2^N$ and the corresponding sequence of Hadamard matrices $H_{4^{N+1}}$. For N even the $2^{2N+1}-2^N-1$ binary codewords of weight $2^{2N}$ together with the null codeword form a Hadamard difference set $v=2^{2(N+1)}$, $k = 2^{2N+1} - 2^N$, $\lambda=2^{2N}-2^N$ and the $2^{2N+1}+2^N$ binary codewords of weight $2^{2N}+2^N$ form a difference set $v = 2^{2N+2}$, $k = 2^{2N+1}+2^N$, $\lambda = 2^{2N}+2^N$. For odd N, the $2^{2N-1}-2^N$ binary codewords each of weight $2^{2N}+2^N$ form a Hadamard difference set $v=2^{2(N+1)}$, $k=2^{2N+1}-2^N$, $\lambda = 2^{2N}-2^N$ and the $2^{2N+1}+2^N-1$ codewords each of weight $2^{2N}$ together with the null codeword form a difference set $v = 2^{2N+2}$, $k = 2^{2N+1}+2^N$, $\lambda = 2^{2N}+2^N$.

These results together with proofs will be reported in another communication.

## References


Ash, R.B. (1966) Information Theory, Interscience Publishers, John Wiley and Sons, New York.

Barlotti, A. (1965) Some Topics in Finite Geometrical Structures, Lecture Notes, Chapel Hill (Inst. of Statistics Mimeo Series no. 439).

Bose, R.C. (1961) On some connections between the design of experiments and information theory. Bull. Intern. Statist. Inst. 38, 257-271.

Bose, R.C. (1963) Some ternary error correcting codes and fractionally replicated designs. Colloques. Inter. CNRS. Paris, no. 110, 21-32.

Bose, R.C. and Mesner, D.M. (1959) On linear associative algebras corresponding to association schemes of partially balanced designs. Ann. Math. Statist. 30, 21-38.

Bose, R.C. and Chakravarti, I.M. (1966) Hermitian varieties in a finite projective space $PG(N,q^2)$. Canad. J. Math. 18, 1161-1182.

Bush, K.A. (1952) Orthogonal arrays of index unity. Ann. Math. Statist. 23, 426-434.

Chakravarti, I.M. (1963) Orthogonal and partially balanced arrays and their application in Design of Experiments, Metrika 7, 231-343.

Chakravarti, I.M. (1971) Some properties and applications of Hermitian varieties in PG(N,$q^2$) in the construction of strongly regular graphs (two-class association schemes) and block designs. Journal of Comb. Theory, Series B, 11(3), 268-283.

Delsarte, P. (1972) Weights of linear codes and strongly regular normed spaces. Discrete Mathematics, 3, 47-64.

Delsarte, P. and Goethals, J.M. (1975) Alternating bilinear forms over GF(q). J. Combin. Theory, 19A,

Dembowski, P. (1968) Finite Geometries, Springer-Verlag 1968.

Goppa, V.D. (1983) Algebraico-geometric codes. Math. USSR Izvestiva, 21(1), 75-91.

Goppa, V.D. (1984) Codes and information. Russian Math. Surveys., 39(1), 87-141.

Hill, R. (1978) Packing problems in Galois geometries over GF(3), Geometriae Dedicata, 7, 363-373.

MacWilliams, F.J. and Sloane, N.J.A. (1977) The Theory of Error-Correcting Codes, North Holland.

Robillard, P. (1969) Some results on the weight distribution of linear codes. IEEE Trans. Info. Theory 15, 706-709.

Raghavarao, D. (1971) Constructions and Combinatorial Problems in Design of Experiements. John Wiley and Sons, Inc., New York.

Segre (1965) Forme e geometrie hermitiane, con particolare riguardo al caso finito. Ann. Math. Pure Appl., 70 1, 202.

Segre, B. (1967) Introduction to Galois Geometries, Atti della Acc. Nazionale dei Lincei, Roma, 8(5), 137-236.

Singleton, R.C. (1964) Maximum distance q-nary codes. IEEE Trans. Info. Theory,10, 116-118.

Tsfasman, M.A. (1982) Goppa codes that are better than the Varshamov-Gilbert bound. Problems of Info. Trans., 18, 163-165.

Tsfasman, M.A., Vladut, S.G. and Zink, T. (1982) Modular curves, Shimura curves and Goppa codes, better than Varshamov-Gilbert bound. Math. Nachr., 104, 13-28.

Vladut, S.G. and Drinfel'd, V.G. (1983) Number of points of an algebraic curve. Functional Anal. Appl. 17, 53-54.

Vladut, S.G., Katsman, G.L. and Tsfasman, M.A. (1984) Modular curve and codes with polynomial complexity of construction. Problems of Info. Transmission 20, 35-42.

Wolfmann, J. (1977) Codes projectifs a deux poids, "caps" complets et ensembles de differences. J. Combin. Theory, 23A, 208-222.

## Résumé

Un bref aperçu des travaux récents de Goppa (1982, 83, 84) est d'abord présenté dans lesquels il a trouvé une construction élégante des codes linéaires à partir des courbes hermitiennes par un recours astucieux au célèbre théorème de Riemann-Roch. Il a obtenu une estimation assez précise du taux de transmission (k/n) et de la capacité (d/n) de correction des erreurs pour ces codes. Ce travail a inspiré Tsfasman, Vladut et Zink (1982) pour construire des codes linéaires $\tilde{q}$-aires a partir des courbes de Shimura définies sur une extension quadratique d'un corps de Galois. Ces codes sont remarquables parce qu'ils franchissent la borne dite de Gilbert-Varshamov pour tout $q = p^{2h} \geq 49$, p un premier.

C'est une extension d'une méthode de construction de Goppa (1983, pp. 76-78) qui nous permet d'abord de plonger une surface hermitienne $x_0^3 + x_1^3 + x_2^3 + x_3^3 = 0$ dans PG(3,4), dans un autre espace PG(9,4) via les séries linéaires des monômes et ensuite d'obtenir de nouveaux codes linéaires ainsi que les répartitions de poids pour ces codes.

Enfin, à partir de la géométrie des intersections d'une surface hermitienne dans $PG(3,s^2)$ par les hyperplans tangents et sécants, nous avons obtenu une famille de codes projectifs linéaires à deux poids. Pour le cas s=2, à partir du graphe fortement régulier associé à ce code, nous avons obtenu l'ensemble (dit de Hadamard) à différences ayant $v=2^8$, $k=2^7-2^3$, $\lambda=2^6-2^3$ et $v=2^8$, $k=2^7+2^3$, $\lambda=2^6+2^3$. En fait, l'auteur vient tout juste démontrer que par une extension de cette méthode on peut construire des ensembles (dits de Hadamard) à différences, ayant les paramètres $v=2^{2N+2}$, $k=2^{2N+1}-2^N$, $\lambda=2^{2N}-2^N$ et $v=2^{2N+2}$, $k=2^{2N+1}+2^N$, $\lambda=2^{2N}+2^N$. Ce dernier résultat fera partie d'un communiqué ultérieur.

---

*Some of these results were presented at the 3ème Colloque International Théorie des Graphes et Combinatoire, Marseille-Luminy, 23-28 Juin 1986 and at the International Conference on Information Processing and Management of Uncertainty in Knowledge-based Systems, Paris, June 30-July 4, 1986.

# Projective Reed-Muller Codes

by

**Gilles LACHAUD**

*J.Équipe CNRS 03 5209*

*Arithmétique & Théorie de l'Information"*

*C.I.R.M.*

*Luminy Case 916*

*13 288 Marseille CEDEX 9*

**Résumé.** On introduit une classe de codes linéaires de la famille des codes de Reed-Muller, les codes de Reed-Muller projectifs. Ces codes sont des extensions des codes de Reed-Muller généralisés ; les codes de Reed-Muller projectifs d'ordre 1 atteignent la borne de Plotkin. On donne les paramètres des codes de Reed-Muller projectifs d'ordre 2.

## 1.Introduction

We note $\mathbf{P}^m(\mathbf{F}_q)$ the projective space of dimension m over the finite field $\mathbf{F}_q$ with q elements, so that

$$\# \mathbf{P}^m(\mathbf{F}_q) = q^m + q^{m-1} + \dots + 1 .$$

We consider the following data :

a) a finite dimensional vector space $\mathcal{L}$ over $\mathbf{F}_q$ ;

b) for every point x of a subset $V \subset \mathbf{P}^m(\mathbf{F}_q)$, an **evaluation map** $\varphi_x : \mathcal{L} \to \mathbf{F}_q$ which is a linear form and associates to every $f \in \mathcal{L}$ an element $\varphi_x(f) \in \mathbf{F}_q$.

With these data, we define the code $C_{\mathcal{L}} \subset \mathbf{F}_q{}^{\#V}$ as the image of the map $\Phi$ from $\mathcal{L}$ to $\mathbf{F}_q{}^{\#V}$ defined by

$$\Phi(f) = (\varphi_x(f))_{x \in V}.$$

The geometric codes are constructed in this way (cf. [8], § 3.1). As a particular case, we can take for $\mathcal{L}$ certain finite dimensional spaces of rational functions on a smooth curve X, and define for a point x of X defined over $\mathbf{F}_q$ which is not a pole of the members of $\mathcal{L}$ the evaluation map as $\varphi_x(f) = f(x)$ ; we then obtain the geometric Goppa codes (cf. [3], [8], [5]). When X is a line, we recover the classical Goppa codes over $\mathbf{F}_q$ as defined in [7].

We will use the above general construction in order to build some codes of the Reed-Muller type, which seems to be more performant than those previously known, as long as the author knows.

## 2. Classical Reed-Muller codes

We now recall the construction of the classical (generalized) Reed-Muller codes in the case $r < q$ after Delsarte, Goethals and Mac Williams (cf. [1]). We consider here the affine space $V = \mathbf{A}^m(\mathbf{F}_q)$ (so that $\# \mathbf{A}^m(\mathbf{F}_q) = q^m$), and choose an integer $r < q$. We take as $\mathcal{L} = \mathbf{F}_q[X_1, \ldots, X_m]_r$ the algebra of **polynomial functions** of degree $\leq r$ with coefficients in $\mathbf{F}_q$ on the affine space $\mathbf{A}^m(\mathbf{F}_q)$. The evaluation map at a point $x \in \mathbf{A}^m(\mathbf{F}_q)$ is just the map $\varphi_x : P \to P(x)$. **The classical (generalized) Reed-Muller code of order** r, which we note $\mathcal{R}_q(r, \mathbf{A}^m)$, is the image of the map $\Phi$ from $\mathbf{F}_q[X_1, \ldots, X_m]_r$ to $\mathbf{F}_q^{\#V}$ defined by

$$\Phi(P) = (P(x))_{x \in V}.$$

The maximal number of zeroes in $\mathbf{A}^m(\mathbf{F}_q)$ of a polynomial of degree r is equal to $rq^{m-1}$ ; we then have :

**Proposition.** *The code* $\mathcal{R}_q(r, \mathbf{A}^m)$ *has length* $n = q^m$, *dimension*

$$k = \binom{r+m}{m},$$

*and minimal distance*

$$d = (q - r)\, q^{m-1}.$$

## 3. Projective Reed-Muller codes

Consider the projective space $\mathbf{P}^m(\mathbf{F}_q)$. There is a canonical projection

$$\pi : \mathbf{A}^{m+1}(\mathbf{F}_q) - \{0\} \quad \to \quad \mathbf{P}^m(\mathbf{F}_q).$$

For $0 \leq i \leq m$, we note $x_i$ the coordinate function of index i :

$$x_i(\lambda_0, \ldots, \lambda_m) = \lambda_i.$$

We note $U_i$ the set $(x_i \neq 0)$ in $\mathbf{A}^{m+1} - \{0\}$, and we set $V_i = \pi(U_i)$ ; the family $\{V_i\}_{0 \leq i \leq r}$ is a covering of $\mathbf{P}^m(\mathbf{F}_q)$. We denote by $\mathbf{F}_q[X_0, \ldots, X_m]^0_r$ the vector space of those $P \in \mathbf{F}_q[X_0, X_1, \ldots, X_m]$ such that P is homogeneous and $\deg(P) = r$.

If $P \in \mathbf{F}_q[X_1, \ldots, X_m]_r$, we define

$$\bar{P}(X_0, X_1, \ldots, X_m) = X_0^r\, P(X_1/X_0, \ldots, X_m/X_0) ;$$

then $\overline{P} \in \mathbf{F}_q[X_0, \dots, X_m]^0_r$ ; the map $P \to \overline{P}$ is an isomorphism from $\mathbf{F}_q[X_1, \dots, X_m]_r$ to $\mathbf{F}_q[X_0, \dots, X_m]^0_r$. We therefore also have

$$\dim \mathbf{F}_q[X_0, \dots, X_m]^0_r = \binom{m+r}{r}.$$

We take $V = \mathbf{P}^m(\mathbf{F}_q)$ ; for $x \in V$, the evaluation map $\varphi_x$ defined on $\mathbf{F}_q[X_0, \dots, X_m]^0_r$ with values in $\mathbf{F}_q$ is defined as follows :

- if $x \in V_o$ : $\varphi_x(P) = \dfrac{P(x)}{x_o^m}$;
- if $x \notin V_o$ but $x \in V_1$ : $\varphi_x(P) = \dfrac{P(x)}{x_1^m}$,
- if $x \notin V_o \cup V_1$ but $x \in V_2$ : $\varphi_x(P) = \dfrac{P(x)}{x_2^m}$,
- etc.

We define as before a map $\Phi$ from $\mathbf{F}_q[X_0, \dots, X_m]^0_r$ to $\mathbf{F}_q^{\#V}$ by

$$\Phi(P) = (\varphi_x(P))_{x \in V}.$$

The map $\Phi$ is an injection on $\mathbf{F}_q[X_0, \dots, X_m]^0_r$ for $r < q$ (as is easily seen by recurrence on m) ; the image of $\Phi$ defines a code $\mathcal{R}_q(r, \mathbf{P}^m) \subset \mathbf{F}_q^{\#V}$.

On the other hand, call $\lambda_r(m)$ the maximal number of zeroes in $\mathbf{P}^m(\mathbf{F}_q)$ of an homogeneous polynomial of degree r ; one has (cf. [6], p. 276) :

$$\lambda_r(m) \leq \frac{r(q^m - 1)}{q - 1}.$$

Gathering together these results,we have proved the following (cf. [8], § 3.2) :

**Theorem.** *Assume* $r < q$. *The code* $\mathcal{R}_q(r, \mathbf{P}^m)$ *has length*

$$n = \frac{q^{m+1} - 1}{q - 1},$$

*dimension*

$$k = \binom{r+m}{r},$$

*and minimal distance*

$$d \geq \frac{q^{m+1} - 1 - r(q^m - 1)}{q - 1}.$$

**Example 1.** When m = 1, the projective Reed-Muller codes are the **generalized Reed-Solomon codes** (cf. [7]). More precisely, we take a set $V \subset \mathbf{P}^1(\mathbf{F}_q)$, with $\# V > r$, and we define as before a map $\Phi$ from $\mathbf{F}_q[X_0, X_1]^0_r$ to $\mathbf{F}_q^{\#V}$ by

$$\Phi(P) = (\varphi_x(P))_{x \in V}.$$

The map $\Phi$ is injective (because # $V > r$), and we then get a code with length $n = \#V$, dimension $k = r + 1$, and minimal distance $d = n - r$ : they are MDS.

**Example 2.** We have proved with M. Martin-Deschamps that every geometric Goppa code can be obtained from a geometric Reed-Muller code by deletion of some of the coordinates.

**4. Projective Reed-Muller Codes of order 1 : The Plotkin bound.**

For $r = 1$, the space $\mathbf{F}_q[X_0, \dots, X_m]_1^0$ is the space of linear forms on $\mathbf{P}^m(\mathbf{F}_q)$ and we have $k = m + 1$ and $d = q^m$ ; for instance the code $\mathcal{R}_2(1, \mathbf{P}^{m-1})$ is equal to the **simplex code** with parameters $[2^m - 1, m, 2^{m-1}]$. For a code C with parameters [n, k, d], the **Plotkin inequality** (cf. [7]) :

$$\# C \leq \frac{d}{d-\theta n} \qquad (\text{when } d > \theta n, \theta = 1 - \frac{1}{q})$$

is equivalent, when $\# C = q^k$, to

$$\frac{d}{n} \leq q^{k-1} \frac{q-1}{q^k - 1}.$$

The right hand side is the **Plotkin bound.**

**Corollary.** *The code* $\mathcal{R}_q(1, \mathbf{P}^m)$ *attains the Plotkin bound.*

**5. Projective Reed-Muller Codes of order 2.**

We obtain more precise results when $q \geq 4$ is even and when $r = 2$. The space $\mathbf{F}_q[X_0, \dots, X_m]_2^0$ is just the vector space of quadratic forms with m+1 variables. The following result can be found in [2] or [4].

**Proposition.** *Let* q *be a power of* 2. *For* $Q \in \mathbf{F}_q[X_0, \dots, X_m]_2^0$ *put* :

$$L_Q = \# \{ x \in \mathbf{P}^m(\mathbf{F}_q) \mid Q(x) = 0 \},$$

$$\lambda_2(m) = \text{Max} \{ L_Q \mid Q \in \mathbf{F}_q[X_0, \dots, X_m]_2^0 \}.$$

*Then*

$$\lambda_2(m) = \frac{q^m - 1}{q - 1} + q^{m-1}.$$

With the help of the preceding proposition we obtain :

**Theorem.** *Assume* q *even. The projective Reed-Muller code* $\mathcal{R}_q(2, \mathbf{P}^m)$ *has length*

$$n = \frac{q^{m+1} - 1}{q - 1},$$

*dimension*

$$k = \frac{(m+1)(m+2)}{2},$$

*and minimal distance* $d = q^{m-1}(q-1)$.

## Bibliography


[1] Delsarte, P., Goethals, J.M., Mac Williams, F.J., *On Generalized Reed-Muller Codes and their Relatives*, Information and Control **16** (1970), 403-442

[2] Dieudonné, J., *La Géométrie des Groupes classiques*, Berlin, Springer 1955.

[3] V.D. Goppa,*Codes and Information*, Uspekhi Mat. Nauk **39** (1984), p. 77–120 ; = Russian Math. surveys **39** (1984), p. 87-141.

[4] Hirschfeld, J. W. P., *Projective Geometries over Finite Fields*, Oxford, Clarendon Press, 1979.

[5] Lachaud, G., *Les codes géométriques de Goppa*, Séminaire Bourbaki 1984/1985, exp. n° 641, Astérisque **133-134** (1986), p. 189-207.

[6] Lidl, R., Niederreiter, H., *Finite Fields*, Enc. of Math. and its Appl., vol. 20, Cambridge University Press, Cambridge, 1983.

[7] MacWilliams, F.J., Sloane, N.J.A., *The Theory of Error-Correcting Codes*, North-Holland, Amsterdam, 1977

[8] Manin, Yu. I., Vladut, S.G., *Codes linéaires et courbes modulaires*, Itogi Nauki i Tekhniki **25** (1984), 209-257 ; = tr. angl. in J. Soviet Math. **30** (1985), 2611-2643 ; = tr. franc. par M. Deza et D. Le Brigand, Pub. Math. Univ. Paris 6, n° 72, 91pp.

# A LOWER BOUND ON THE MINIMUM EUCLIDEAN DISTANCE OF TRELLIS CODES

*Marc Rouanne*

*Daniel J. Costello, Jr.*

*Dept. of Elec. & Comp. Engr.*

*Univ. of Notre Dame*

*Notre Dame, IN 46556*

Submitted to

"Trois journées sur le codage"

**Abstract: A lower bound on the minimum Euclidean distance of trellis codes is considered. The bound is based upon Costello's free distance bound for convolutional codes [1]. The bound is a random coding bound over the ensemble of non-linear time-varying Euclidean trellis codes. We compare schemes using different signal constellations and mappings and apply the bound to particular trellis coded modulation (TCM) schemes such as Ungerboeck's [3] and Lafanechere and Costello's [4].**

*Keywords:* Trellis-coded modulation, minimum Euclidean distance, random coding bound.

---

This work was supported by NASA grant NAG5-557 and NSF grant ECS84-14608.

*Mots-clés:* Codes de treillis et modulation, distance Euclidiennne minimale, borne minorante.

## 1 INTRODUCTION

A TCM scheme is defined by a binary trellis encoder, a signal constellation, and a mapping of signals onto the trellis. Let $k$ be the information block length and $\nu$ the memory length of the encoder. Then $\nu_0 = k\,\nu$ is the constraint length of the code (we assume that the encoder contains $k$ memory registers of equal length $\nu$). The encoder is a finite state machine with $2^{\nu_0}$ states and $2^k$ branches to and from each state. The error probability of a code used with maximum-likelihood (Viterbi) decoding on an AWGN channel can be bounded in terms of its free distance $d_{free}$. Therefore an efficient mapping will assign channel signals to branches to achieve maximum free distance $\max(d_{free})$. The lower bound on $\max(d_{free})$ uses a random coding argument based on the moment generating function $\overline{e^{\alpha\rho d^2} \sum_{\underline{y}} \left[ \sum_{\underline{y}'} e^{-\alpha d_e^2(\underline{y},\underline{y}')} \right]^{\rho}}$, where the overbar indicates an average over the ensemble of all code choices (for a given trellis), the first sum is over all choices of paths $\underline{y}$, the second sum is over all paths $\underline{y}'$ diverging from $\underline{y}$, $\alpha$ and $\rho$ are arbitrary constants, $d$ is the lower bound on $\max(d_{free})$, and $d_e(\underline{y},\underline{y}')$ is the Euclidean distance between paths $\underline{y}$ and $\underline{y}'$.

## 2 DERIVATION OF THE BOUND

Forney [2] set forth clearly in six steps a bounding technique which owes much to Chernoff [7], Gallager [8], and Viterbi [9]. We use only the first four steps.

### 2.1 Step 1: Define a random ensemble

The random ensemble is the set of all "non-linear, time-varying codes" associated with a specific signal constellation and a specific trellis.

## 2.2 Step 2: Obtain a bound of the moment-generating function form:

If the probability over all codes in the random ensemble that $d_{free}$ is smaller than some $d$ is less than 1, there must exist at least one code for which $d_{free}$ is actually larger than or equal to $d$. This can be written as follows:

$$\underset{\text{all codes}}{P}\,(d_{free}(code) \; < \; d) \; < \; 1 \;\rightarrow\; \textit{there exists a code such that } d_{free} \geq d.$$

Furthermore,

$$d_{free}(code) \;\geq\; d \quad \text{if} \quad T_{\alpha,\rho}(code,d) \;<\; 1,$$

where
$$T_{\alpha,\rho}(code,d) \;\equiv\; e^{\alpha\rho d^2} \sum_{\underline{y}} \left[ \sum_{\underline{y}'} e^{-\alpha d_e^2(\underline{y},\underline{y}')} \right]^{\rho}.$$

Hence,

$$\textit{there exists a code such that } d_{free} \geq d \;\text{ if }\; \overline{e^{\alpha\rho d^2} \sum_{\underline{y}} \left[ \sum_{\underline{y}'} e^{-\alpha d_e^2(\underline{y},\underline{y}')} \right]^{\rho}} \;<\; 1. \tag{1}$$

## 2.3 Step 3: Configuration counting

Equation (1) contains an average over all codes of a function $T_{\alpha,\rho}$, where $T_{\alpha,\rho}$ depends on the codewords of a *code*. Shannon's random coding technique switches these two operations and obtains an average over all codewords of another function $\hat{T}_{\alpha,\rho}$, where $\hat{T}_{\alpha,\rho}$ is an average over all the codes (it can be viewed as the probability that a codeword is chosen). Here, the codewords are the paths through the trellis. "Configuration counting" determines the number of occurrences of a path in all the codes. This enables us to calculate $\hat{T}_{\alpha,\rho}$.

## 2.4 Step 4: Reduce the above bound to a usable form (Gallager's Lemma)

Step 4 (see Gallager [8]) puts the bound in its final form.

# 3 STATEMENT OF THE BOUND

Given a signal constellation S, there exists a $(k, v_0)$ trellis code with minimum Euclidean distance $d_{free}$ such that:

$$d_{free}^2 \geq \max_{\begin{cases} E(\alpha,\rho) > \ln 2 \\ 0 \leq \alpha \\ 0 \leq \rho \leq 1 \\ p(s) \end{cases}} \left[ v_0 \frac{E(\alpha\,,\rho)}{\alpha} - \frac{F[\,E(\alpha\,,\rho)\,]}{\alpha} \right], \tag{2}$$

where

$\alpha$ and $\rho$ are parameters that enable us to optimize the bound,

$s$ and $s'$ are signal points from S,

between $s$ and $s'$,

$p(s)$ is a probability distribution on S, and

$$E(\alpha,\rho) = -\ln \left[ \sum_{s\in S} p(s) \left[ \sum_{s'\in S} p(s')\, e^{-\alpha\, d_e^2(s\,,s')} \right]^{\rho} \right]^{\frac{1}{\rho k}} .$$

# 4 USING THE BOUND TO EVALUATE TCM SCHEMES

## 4.1 $d_{free}$ viewed as a function of $v_0$:

The lower bound on $d_{free}$ increases with the constraint length $v_0$. Bounds on the free Hamming distance of binary convolutional codes have usually been expressed as functions of $v_0$. This leads to asymptotic, exponentially tight bounds on $\max(d_{free})$ for large $v_0$, but the bounds are not tight for small constraint lengths. In the following paragraphs we compare constellations at a given rate $R$ of information bits per dimension and at a given average energy $E_{avg}$ per dimension.

### 4.1.1 The effect of increasing the number $M$ of signals on $\max(d_{free})$ (same rate $R$, same dimensionality, same average energy $E_{avg}$):

Increasing the number $M$ of signals increases $\max(d_{free})$ because it provides more flexibility in the mapping of signals to trellis branches. In particular, this flexibility generates some codes with larger distances between correct and incorrect paths than codes generated with fewer signals. Note that increasing $M$, while keeping $k$ constant, lowers the code rate $R_c \equiv \frac{k}{\log_2 M}$.

### 4.1.2 The effect of increasing the signal set dimensionality on $\max(d_{free})$ (same rate $R$, same number of signals $M$, same average energy $E_{avg}$):

Increasing the signal set dimensionality increases $\max(d_{free})$. We obtain higher dimensional schemes by using a basic one- or two-dimensional constellation for several transmission intervals [4]. Fig.1 shows an example of an L-dimensional scheme (L uses of 4-PAM, L=1,2,3, and 4). For a given constraint length $\nu_0$, higher dimensionality yields a larger $\max(d_{free})$, and as $\nu_0$ increases the curves diverge and the gain goes to infinity. However, the error coefficient increases with dimensionality [10], which may cancel the gain from the increased $d_{free}$ at moderate decoded bit error rates.

### 4.1.3 The effect of changing the modulation scheme on $d_{free}$ (same dimensionality, same number of signals $M$, same average energy $E_{avg}$, same rate $R$):

Changing the modulation scheme affects $\max(d_{free})$. For example, rectangular constellations yield a larger $\max(d_{free})$ than constant envelope constellations. Fig.2 shows M-PSK vs. M-QASK, and suggests that a $d_{free}$ penalty is associated with constant envelope modulation.

### 4.1.4 The effect of uncoded bits or parallel transitions on $\max(d_{free})$ when the signal constellation is "mapped by set partitioning":

Uncoded bits generate parallel transitions in the trellis and limit $\max(d_{free})$, since $d_{free}$ cannot be larger than the minimum distance between parallel transitions (the minimum distance within the subsets created by "set partitioning"). Fig.3 shows that for 8-PSK, 1 uncoded bit increases $d_{free}$ for small constraint lengths. Although the bound is not tight, its slope agrees with the best known codes constructed [3]. For large $\nu_0$, we see that 1 uncoded bit limits the achievable $d_{free}$.

## 4.2 Comparison of the lower bound on $\max(d_{free})$ with an upper bound [5,6] and known codes [3]:

We expect upper bounds to be tighter than our lower bound for small constraint lengths, and our lower bound to be tighter for large constraint lengths (this is analogous to the way bounds on binary convolutional codes behave). The lower bound is a "Chernoff type" bound and therefore is exponentially tight when $\nu_0$ is large. Fig.4 shows that the upper bound is tighter than the lower bound for small constraint lengths. But the lower bound becomes tighter as $\nu_0$ increases. The slope of the lower bound gives a precise indication of the asymptotic rate of increase in $\max(d_{free})$. Finally, the lower bound guarantees the existence of codes that can achieve a certain $d_{free}$.

# 5 CONCLUSION

We have obtained a random coding lower bound on $\max(d_{free})$. It takes into account the actual Euclidean distances between points in the channel signal constellation. As with any "Gilbert" bound, it is most useful as an asymptotic bound. It provides a means of comparing the asymptotic performance of different modulation schemes, and proves the existence of codes that achieve $d_{free}$ larger than the bound.

## REFERENCES


[1] D. J. Costello, Jr., "Free Distance Bounds for Convolutional Codes", IEEE Trans. Inform. Theory, vol. IT-20, pp. 356-365, May 1974.

[2] G. D. Forney, Jr., "Convolutional Codes II: Maximum Likelihood Decoding", Informa- tion and Control , vol. 25, pp. 222-266, July 1974.

[3] G. Ungerboeck, "Channel Coding with Multilevel Phase Signals", IEEE Trans. Inform. Theory, vol. IT-28, pp. 55-67, Jan. 1982.

[4] A. Lafanechere and D. J. Costello, Jr., "Multidimensional Coded PSK Systems Using Unit-Memory Trellis Codes", Proceedings Allerton Conf. on Communication, Control, and Computing, pp. 428-429, Monticello, IL, Sept. 1985.

[5] A. R. Calderbank, J. E. Mazo, and V. K. Wei, "Asymptotic Upper Bounds on the Minimum Distance of Trellis Codes", IEEE Trans. Commun. , vol. COM-33, no. 4, pp. 305-310, April 1985.

[6] A. R. Calderbank, J. E. Mazo, and H. M. Shapiro, "Upper Bounds on the Minimum Distance of Trellis Codes", Bell Syst. Tech. J., vol. 62, pp. 2617-2646, Oct. 1983, part I.

[7] H. Chernoff, "A Measure of Asymptotic Efficiency for Tests of a Hypothesis Based on a Sum of Observations," Ann. Math. Stat. , vol. 23, pp. 493-507, 1952.

[8] R. G. Gallager, "A simple Derivation of the Coding Theorem and Some Applications," IEEE Trans. Inform. Theory , vol. IT-11, pp. 3-18, Jan. 1965.

[9] A. J. Viterbi, "Error Bounds for Convolutional Codes and an Asymptotically Optimum Decoding Algorithm," IEEE Trans. Inform. Theory , vol. IT-13, pp. 260-269, Apr. 1967.

[10] G. D. Forney, Jr., R. G. Gallager, G. R. Lang, F. M. Longstaff, and S. U. Qureshi, "Efficient Modulation for Band-limited Channels", IEEE J. on Selected Areas in Com- munication , vol. SAC-2, no. 5, pp. 632-647, Sept. 1984.

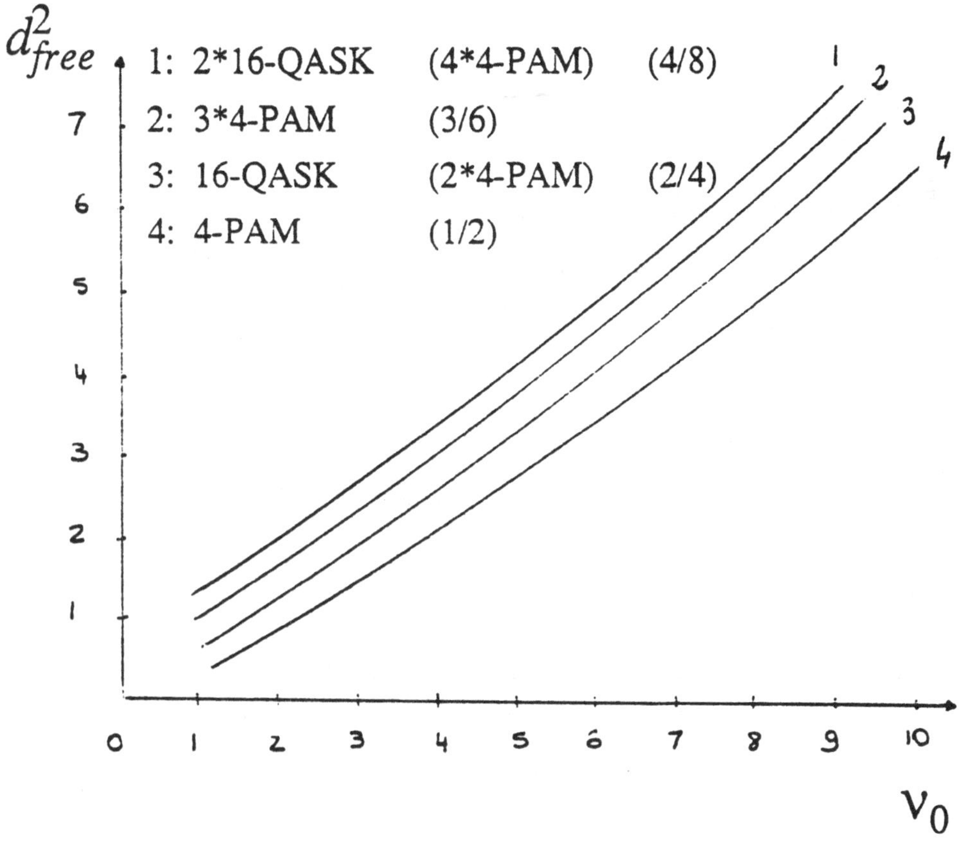


Fig. 1: Comparison of schemes with different dimensionalities, $R = 1 bit/dim$, $E_{avg} = 1$.

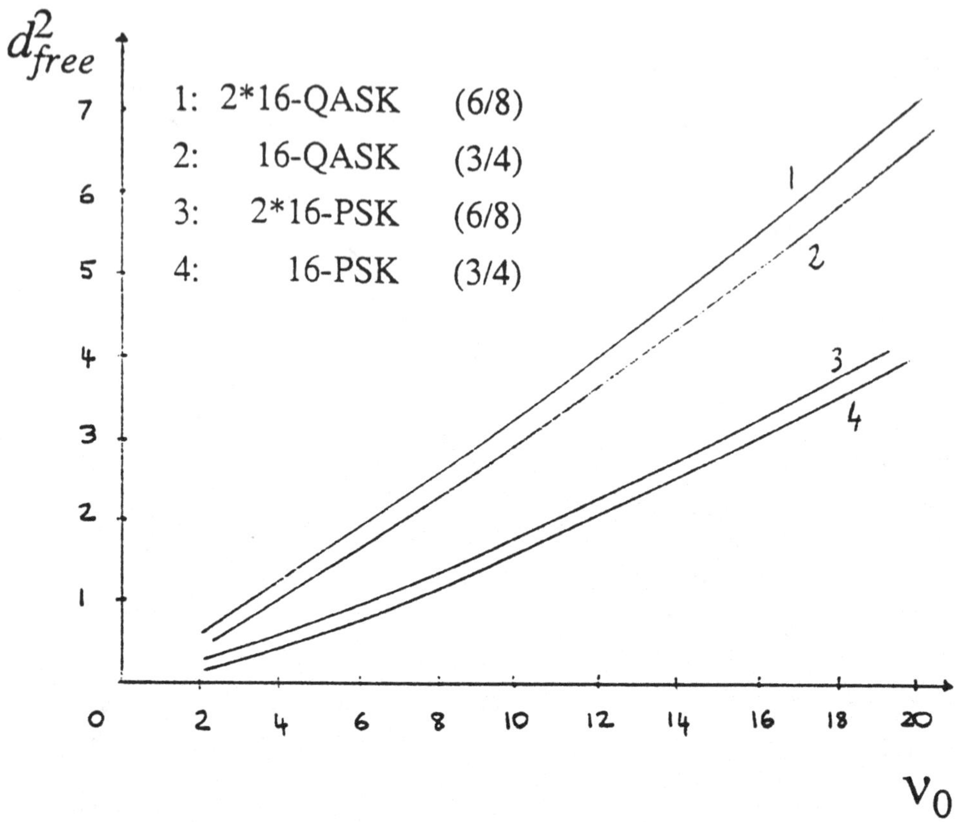


Fig. 2: Comparison of M-PSK, M-QASK, L*M-PSK and L*M-QASK schemes, $R = 1.5$ *bit/dim*, $E_{avg} = 1$.

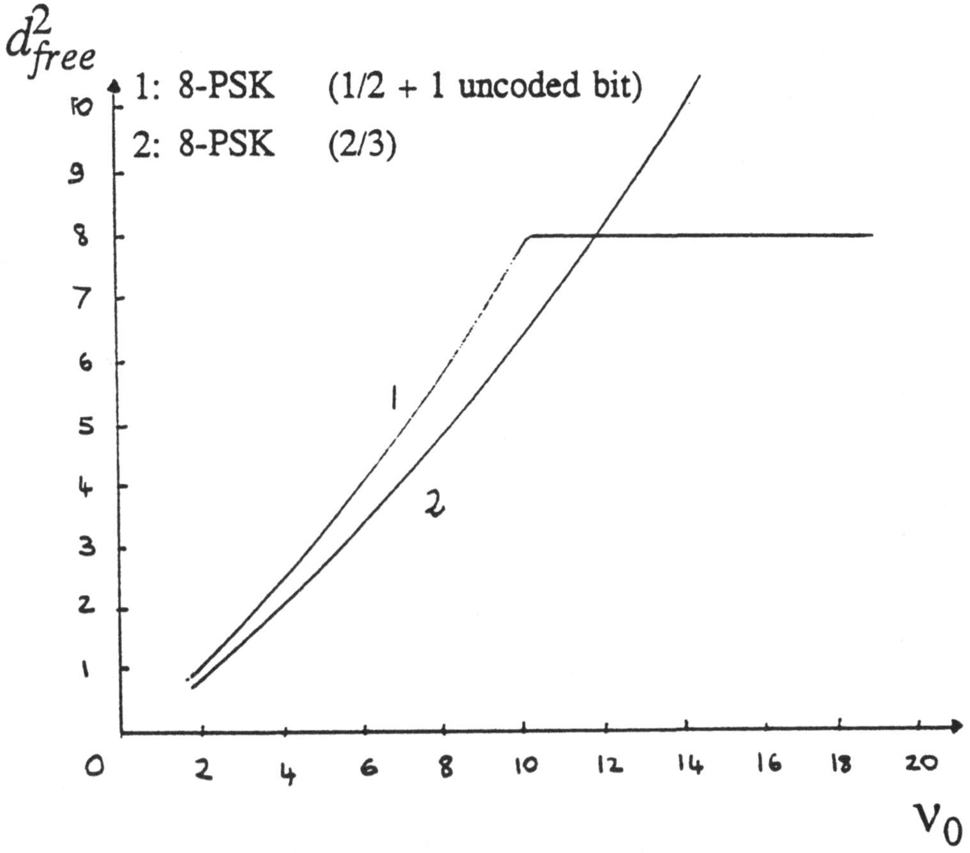


Fig. 3: Trellis coding and uncoded bits (Ungerboeck) for 8-PSK. $R = 1$ *bit/dim*, $E_{avg} = 1$.

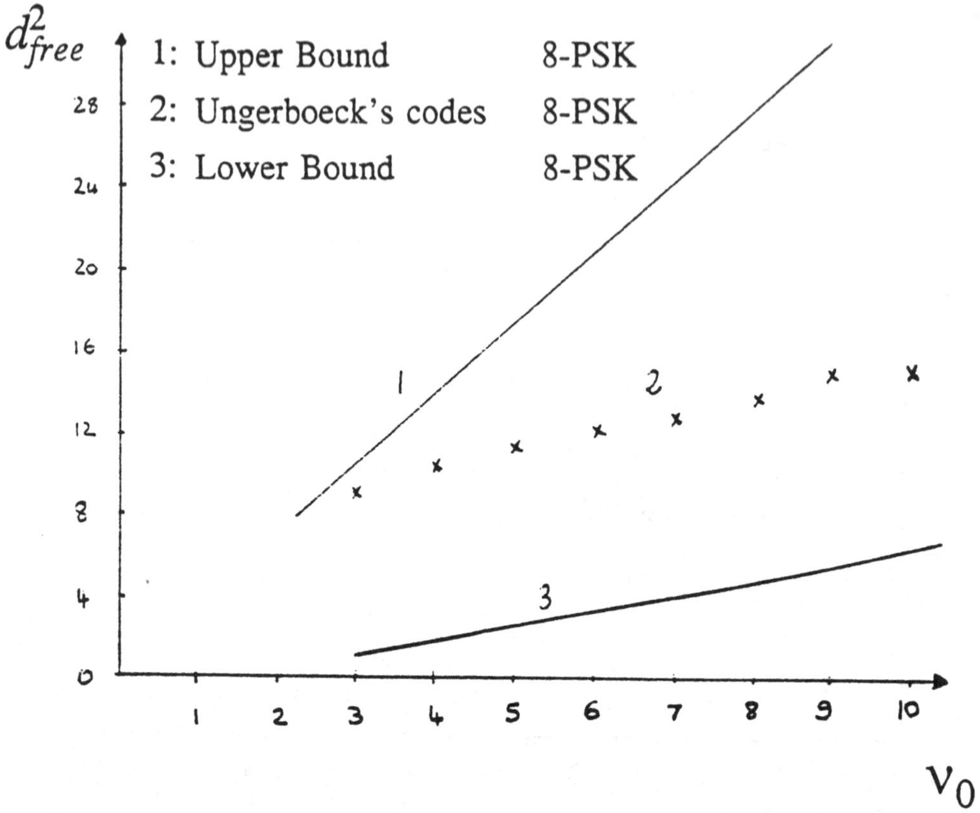


Fig. 4: Comparison of the lower bound on $d^2_{free}$ with upper bounds (Calderbank, Mazo, and Wei).

# On the Inherent Intractability of Soft Decision Decoding of Linear Codes

J. FANG, G. COHEN, P. GODLEWSKI, G. BATTAIL
Département Systèmes et Communications, CNRS UA 820,
Ecole Nationale Supérieure des Télécommunications
46, Rue Barrault, 75634, Paris, Cédex 13, France

## 1. Introduction

This note was inspired by an article of Berlekamp et al. /1/, in which they proved that the general hard decision decoding of linear codes is an inherently intractable problem in the sense of the theory of NP-completeness /4/, i.e., no algorithm running in polynomial time can exist for this problem unless P=NP.

In this work, we show, on the basis of the work mentioned above, that the general soft decision decoding problem can be formulated, without loss of generality, in a form such that it can be seen as a generalization of the hard decision decoding problem. Then it is shown to be also an NP-complete problem, and no algorithm in polynomial time exists for it, unless P=NP.

Another interesting problem studied in /1/ is the computational complexity of finding the minimum distance of linear codes. This problem, conjectured by the authors of /1/ to be NP-complete, has been open up to date. Several authors have reported some results on the related problems, in /2/ and more recently in /3/. The authors of /3/ show some relation of this problem to the decoding problem. These results provide further evidence to support this conjecture. We discuss briefly this point in the conclusion.

## 2. Hard Decision Decoding as NP-Complete Problem

The hard decision complete decoding (HDCD) problem is the following optimization problem:

Instance: H, an (n-k)xn binary matrix, and
$\underline{r}$ , a binary n-vector.

(HDCD)

Question: Find $\underline{c}^*$, $\underline{c}^*H^t = \underline{0}$, such that

$$D_H(\underline{c}^*,\underline{r}) = \min_{\underline{c}H^t=\underline{0}} D_H(\underline{c},\underline{r}) .$$

where $D_H$ is Hamming distance. It is called also the minimum distance decoding problem, since it requires to find the codeword closest to the received vector. To this problem we associate its decision version ("yes-no" problem) Hard Decision Decoding (HDD) as follows, which enables us to analyse the problem using the NP-completeness theory:

(HDD)

Instance: H, an $(n-k) \times n$ binary matrix,
$\underline{r}$ , a binary n-vector, and a positive integer w.

Question: Is there $\underline{c}^*, \underline{c}^* H^t = \underline{0}$, such that
$D_H(\underline{c}^*, \underline{r}) \leq w$ ?

We can see that HDCD is not easier than HDD, since we can solve HDD if we can solve HDCD by comparing the solution of HDCD with w. HDD is also called bounded distance decoding, and in practice w is usually assigned to be $t = \lfloor (d-1)/2 \rfloor$ , as in the case of the algebraic decoding (where d is the minimum distance of the code characterized by H).

We state first a lemma:

Lemma: HDD is equivalent to CW(Coset Weight).

CW is the following decision problem:

(CW)

Instance: H, an $(n-k) \times n$ binary matrix, $\underline{s}$ a binary $(n-k)$-vector, and a positive integer w.

Question: Is there $\underline{e}, \underline{e}H^t = \underline{s}$, such that $wt(\underline{e}) \leq w$?

This lemma is shown as follows:

Take any instance (H, $\underline{s}$ , w) of CW, and construct (in polynomial time) (H, $\underline{r}$, w), by computing $\underline{r}$ from the equation $\underline{x}H^t = \underline{s}$. Now, if there exists $\underline{c}^*$ such that $\underline{c}^* H^t = \underline{0}$ and $wt(\underline{c}^* \oplus \underline{r}) \leq w$ (answer YES to HDD ), set $\underline{e} = \underline{r} \oplus \underline{c}^*$: $\underline{e}$ is a solution to CW: $\underline{e}H^t = \underline{r}H^t \oplus \underline{c}^* H^t = \underline{s}$. Conversely, if there is $\underline{e}$ such that $\underline{e}H^t = \underline{s}$ and $wt(\underline{e}) \leq w$ ( answer YES to CW ), set $\underline{c}^* = \underline{r} \oplus \underline{e}$ , and $\underline{c}^*$ is a solution to HDD. So a polynomial time algorithm for HDD would solve CW in polynomial time.

In /1/, CW has been proven to be an NP-complete problem. From above, CW is polynomially reduced to HDD, and it is evident that HDD $\in$NP , so HDD $\in$NP-C. Consequently, HDCD is an NP-hard problem, which implies the inherent intractability of the general hard decision decoding problem.

## 3. Soft Decision Decoding Problem

The soft decision decoding makes use of the channel reliability information of the received symbols, providing better performance than hard decision. We study the soft decision decoding problem for the discrete memoryless channel (DMC). A DMC is the triple: $\left\{J, Q, P_r(Q/J)\right\}$, where J is the input symbol set, Q is the output symbol set, $P_r(Q/J)$ is the transition probability matrix. By soft decision we mean that the size of Q is larger that that of J. We suppose that the entries in the transition probability matrix are rationals, and encodable in finite length binary strings. This problem will be treated in the appendix, where we show that these coefficients satisfy this property when the DMC is obtained by quantizing an additive gaussian white noise channel. We define and utilize in the sequel the decoding metrics as: $\mu(i,j)=f(\log Pr(i/j))$, for $i \varepsilon Q, j \varepsilon J$, where f may be any monotonic increasing function that maps the quantities logPr(i/j) into positive integers. If the metric word length is sufficiently long, the error caused by this treatment can be ignored. We define also $\mu(\underline{r},\underline{c})=\sum_{i=1}^{n}\mu(r_i,c_i)$.

Without loss of generality, the soft decision decoding of linear codes we consider is the so called maximum likelihood decoding (MLD), which can be formulated in terms of a DMC as follows:

Instance: H, an (n-k)xn binary matrix, decoding metrics
$\mu(i,j)$, integers, ( $i \varepsilon Q$, Q is output alphabet,
and $j \varepsilon J$, J is the input alphabet of the DMC).
$\underline{r} \varepsilon Q^n$: the received vector.

(MLD)

Question: Find $\underline{c}^*$, $\underline{c}^* H^t = \underline{0}$ , such that
$$\mu(\underline{r},\underline{c}^*) = \underset{\underline{c}H^t=\underline{0}}{\text{Max}}\, \mu(\underline{r},\underline{c}).$$

Now we transform the formulation to construct the problem as a generalization of HDCD. For a DMC, we have

$$\underset{\underline{c}\varepsilon C}{\text{Max}}\, \mu(\underline{r},\underline{c}) \rightarrow \underset{\underline{c}\varepsilon C}{\text{Max}} \sum_{i=1}^{n} \log P_r(r_i/c_i)$$

$$\rightarrow \underset{\underline{c}\varepsilon C}{\text{Max}} \sum_{i=1}^{n} (\log Pr(r_i/c_i) - \log Pr(r_i/hd(r_i)))$$

where $hd(r_i)$ is the hard decision on $r_i$, defined by:

$hd(r_i) \varepsilon J$, and $Pr(r_i/hd(r_i)) = \underset{j\varepsilon J}{\text{Max}}\, Pr(r_i/j)$ .

So we have

$$0 < Pr(r_i/c_i)/Pr(r_i/hd(r_i)) \leq 1$$

$$0 \leq -\log\ (Pr(r_i/c_i)/Pr(r_i/hd(r_i))) < \infty .$$

Define the likelihood coefficient

$$v_i = -\log(Pr(r_i/c_i)/Pr(r_i/hd(r_i))) = \begin{cases} 0, & \text{if } c_i = hd(r_i), \\ \geq 0, & \text{otherwise.} \end{cases}$$

Then we can write

$$\underset{\underline{c}\in C}{\mathrm{Max}}\ \mu(\underline{r},\underline{c}) \rightarrow \underset{\underline{c}\in C}{\mathrm{Min}} \sum_{i=1}^{n} -\log\ Pr(r_i/c_i)/Pr(r_i/hd(r_i))$$

$$= \underset{\underline{c}\in C}{\mathrm{Min}} \sum_{i=1}^{n} D_H(c_i, hd(r_i))v_i = \underset{\underline{c}\in C}{\mathrm{Min}}\ D_G(\underline{c},\underline{r})$$

where $D_G$ is the generalized distance:

$$D_G(c_i, r_i) = D_H(c_i, hd(r_i))v_i .$$

The concept of the generalized distance was firstly introduced by Forney, /5/, and this one was used as the "optimal weighting coefficient of symbol" for a trellis soft decision decoding algorithm by Battail /6/. In the appendix, the coefficients $v_i$ are proved to be representable by finite length binary strings, so they will be treated equivalently as integers in the sequel.

We state now the general soft decision decoding problem formulated as the following "Soft Decision Complete Decoding" (SDCD) problem:

(SDCD)

Instance: H, (n-k)xn binary matrix, $\underline{r} \in Q^n$, received vector, $\underline{v}$, integer vector of likelihood coefficients.

Question: Find $\underline{c}^*$, $\underline{c}^* H^t = \underline{0}$, such that

$$D_G(\underline{c}^*,\underline{r}) = \underset{\underline{c}\in C}{\mathrm{Min}}\ D_G(\underline{c},\underline{r}).$$

The associated decision problem is the following:

(SDD)

Instance: H, $\underline{r}$, $\underline{v}$, a positive integer w.

Question: Is there $\underline{c}$, $\underline{c}H^t = \underline{0}$, such that

$$D_G(\underline{c},\underline{r}) \leq w \ ?$$

We have the following result as a corollary of that about CW in /1/:

Theorem: SDD is NP-complete.

Proof: The proof of this is straightforword. Note that HDD is a degenerate case of SDD in which the symbol likelihoods are all equal, i.e., we may take $\underline{v}=\underline{1}$, the all 1's vector, we take also $hd(r_i)=r_i$, for $i=1,\dots,n$, then $D_G(\underline{c},\underline{r}) = D_H(\underline{c},\underline{r})$. So any hypothetical polynomial time algorithm for SDD could solve HDD in polynomial time. The NP-completeness of SDD follows, because SDD $\varepsilon$ NP. The maximum likelihood decoding in the form of SDCD, is then NP-hard, and no algorithm in polynomial time exists for soft decision decoding of linear codes, unless P=NP.

## 4. Conclusion and Discussion

The NP-completeness of general soft decision decoding of linear codes has been shown.

The soft decision decoding can be formulated as a generalized form of the hard decision decoding by defining the generalized distance, which is an analogous function of the channel outputs. This brings some additional problems such as quantization, word length effect, hybrid computation, etc. These problems are difficult to analyse, but easy to solve in practice, cf. for example, /8/. For our complexity analysis need, we give an analysis of the decoding metric word length problem in the appendix.

The syndrome is not defined for soft decision schemes, so an equivalent form of hard decision decoding is then introduced to overcome the difficulty in analysing the complexity. Let us examine the following formulation of soft decoding in terms of the syndrome:

Instance: H, $\underline{s},\underline{v}$, a positive integer w.
Question: Is there $\underline{e},\underline{e}H^t = \underline{s}$, such that
$D_G(\underline{0},\underline{e}) \leq w$ ?

It is evidently an NP-complete problem, but it is not a good formulation of the soft decision decoding. The reason for this is that the coset leader corresponding to the syndrome does not always give the error pattern that makes the analogous metric $\underline{ev}^t$ minimum. This fact shows the difficulty in the attempt of algebraic approaches for decoding with analogous metrics.

In /3/, the decoding problem is shown being related to the problem of finding the minimum distance of linear codes. The NP-completeness of related problems is shown in /3/, with the following conditions:

- the minimum weight codeword searched for has "1" as its first component;
- the minimum weight codeword searched for has a fraction of p/(p+1) of its "1" on its first components.

The NP-completeness of the first problem is shown by reducing CW (Coset Weight) to it. It is easy to see that the second result can be used to show the NP-Completeness of the following problem $CW_p$ , related to CW:

Instance: $H$, $\underline{s}$, $w \in N$, $p \geq 3$, $p \in N$
Question: Is there $\underline{e}$, $\underline{e}H^t = \underline{s}$, $wt(\underline{e}) \leq w$, and $e_1 = e_2 = \cdots = e_{\lfloor wp/(p+1) \rfloor} = 1$ ?

since the second one of the above NP-complete problems is just a subproblem of $CW_p$ with the restriction of $\underline{s}$ being $\underline{0}$ .

So we have the implication for the decoding problem that the decoding remains intractable even if a fraction p/(p+1) of the errors could be guessed, provided that the number of remaining errors is unknown. The problem can again be restated without use of the syndrome and generalized to the soft decision case.

**Acknowledgment**

The authers would like to thank an anonymous referee for reading the original version very carefully. This referee indicated some errors and made several very useful suggestions regarding the presentation of this paper.

## Appendix: Word Length for Representing the Decoding Metrics

The main operation in most known maximum likelihood decoding algorithms such as Viterbi decoding, sequential decoding as well as most block code soft decoding algorithms, is the comparison of the word (or path) metrics. This is even seen in the decoding criterion we used in section 3, from which it seems to us that processings more complicated than computation of decoding metrics and comparison are not necessary. The following criterion is equivalent to the (SDCD) problem:

$$\hat{\underline{c}} = \underline{c}^*, \quad \text{such that } \forall\, \underline{c} \in C\,, \quad \underline{c} \neq \underline{c}^*, \quad D_G(\underline{c}^*, \underline{r}) \leq D_G(\underline{c}, \underline{r}),$$

where C is the code considered. The generalized distance $D_G$ is in fact the intercorrelation between the weighting coefficient vector and the hard decision vector. The computation of a word metric simply consists of the addition of the symbol metrics. If the largest weighting coefficient is $V_{max}$, then $Sup(D_G) = n\,V_{max}$, where n is the block or path length.

In addition, decoding is usually performed by making comparisons between the word metrics or path metrics. A comparison is equivalent to a subtraction which does not increase the word length of the concerned metrics. Therefore, we can think that most soft decision decoding algorithms can be implemented by fixed point operation, as long as the word metrics are of finite word length.

Now we show this point for the word metric defined earlier:

$$V_i = -\log(P_r(r_i/c_i)/P_r(r_i/hd(r_i))).$$

Suppose we have a binary input, uniformely quantized Q-output DMC, and a zero is assumed to have been transmitted, on the AGWN unquantized channel.

$$Sup(V_i) = -\log(P_r(J/-1)/P_r(J/1)) \quad ,$$

where $J = \sqrt{E_s}$ is the largest output of the DMC, $E_s$ is the energy of channel symbol. Let L denote the number of quantization levels, $\Delta$ the quantization interval, equal to $\Delta = 2\sqrt{E_s}/L$ and $p(r_i/c_i)$ the transition probability densities of the unquantized AGWN channel. The unquantized channel output $r_i$ is represented by J iff $r_i > y = \sqrt{E_s} - \Delta/2 = \sqrt{E_s}(L-1)/L$. We thus have

$$p_r(\sqrt{E_s}/1) = \int_y^{\infty} p(x/1)dx = Q(\alpha_1),$$

$$p_r(\sqrt{E_s}/-1) = \int_y^{\infty} p(x/-1)dx = Q(\alpha_2),$$

$$\text{with} \quad \alpha_1 = -\sqrt{2\gamma_s}/L \;, \alpha_2 = \sqrt{2\gamma_s}(2 - 1/L) \;,$$

where $\gamma_s$ is the ratio of signal noise per channel symbol, and

the Q(.) function being defined as $Q(x) = \int_x^{\infty} \frac{1}{\sqrt{2\pi}} e^{\frac{-t^2}{2}} dt.$

When $\gamma_s \gg 1$, $Q(\alpha_1) \sim 1/2+\varepsilon, 0<\varepsilon<1/2$, $Q(\alpha_2) \sim \frac{1}{\sqrt{2\pi}\alpha_2} e^{\frac{-\alpha_2^2}{2}}$,

$$\mathrm{Sup}(V_i) \sim \log \alpha_2 + \frac{\alpha_2^2}{2} + c_1, \quad \text{where} \quad c_1 = \left(\frac{\log 2 + \log \pi}{2}\right)$$

$$\mathrm{Sup}(V_i) \sim \gamma_s (2 - 1/L)^2 + \frac{1}{2}\log\gamma_s + \log(2 - 1/L) + c_2$$

$$\text{where} \qquad c_2 = \log 2 + \frac{1}{2}\log\pi \,,$$

$$\mathrm{Sup}(V_i) \sim \gamma_s \left[(2 - \frac{1}{L})^2 + \frac{1}{2}\log\gamma_s/\gamma_s + \log(2 - \frac{1}{L})/\gamma_s + c_2/\gamma_s\right]$$

$$\sim \gamma_s (2 - \frac{1}{L})^2 \,, \qquad \text{when} \quad \gamma_s \gg 1$$

The number of bits for representing a word metric is nearly:

$$M = \log_2 \left\{ n\mathrm{Sup}(v_i) \right\} \leq \lceil \log_2 n \rceil + \lceil \log_2 \gamma_s \rceil + \lceil 2\log_2 (2 - \frac{1}{L}) \rceil$$

We know that the signal to noise ratio $\gamma_s$ is physically limited, so M must be a finite quantity. The number of quantization levels contributes little to the word length of the metric so defined.

A similar result holds for the other decoding metric definitions such as Fano's metric /7/, the suboptimum quantized metric used in Viterbi decoding /8/, since they are all linear functions of logarithms of the channel transition probabilities.

In summary, the decoding metrics can be encoded with finite word length, and the comparison, which does not increase the word length, is the only real number operation for most soft decision decoding algorithms. The fixed point computation suffices for their implementation. The fixed point word operation is equivalent to finite length integers operation in the complexity analysis and in the special purpose signal processors. This shows that it is reasonable to treat the weighting coefficients as integers in our soft decision decoding problem formulation.

## References

/1/. E.R. Berlekamp, R.J. McEliece and H.C.A. van Tilborg, "On the Inherent Intractability of Certain Coding Problems", IEEE, Trans. on IT, Vol. IT-24, No.3, May, 1978.

/2/. S.C. Ntafos, S.L. Hakimi, "On the Complexity of Some Coding Problems", IEEE Trans. on IT, Vol. IT-27, No.6, November 1981.

/3/. A. Lobstein, G. Cohen, "Sur la Complexité d'un Problème de Codage" , RAIRO on Theoretical Informatics and Applications, Vol. 21, No. 1, 1987. Gauthier-Villars, Paris.

/4/. M. R. Garey and D. S. Johnson, "Computers and Intractability: A Guide to the Theory of NP-completeness", San Francisco: Freeman, 1975.

/5/. G.D.Jr. Forney, "Concatenated Codes", MIT press. Cambridge, May, 1966.

/6/. G. Battail, "Décodage Pondéré Optimal des Codes Linéaires en Blocs I - Emploi Simplifié du Diagramme du Treillis", Annales des Télécommunications, 38, 11-12, pp.443-459.

/7/. R. M. Fano, "A Heuristic Discussion of Probabilistic Decoding", IEEE, Trans. IT-9. pp.66-74, Apr. 1963.

/8/. A.J. Viterbi, J.K. Omura, "Principles of Digital Communication and Coding" , McGraw-Hill, 1971.

# WEIGHTED DECODING AS A MEANS FOR REESTIMATING A PROBABILITY DISTRIBUTION (ABSTRACT)

Gérard Battail,
Ecole Nationale Supérieure des Télécommunications,
46, rue Barrault, F - 75634 PARIS CEDEX 13, France

## 1 - A REDEFINITION OF DECODING

We propose to redefine weighted decoding of a redundant code as consisting of reestimating a given prior probability distribution, available for each received symbol as the demodulator output, in order to take into account the code constraints. The result from such redefined decoding will be referred to as posterior probability distribution.

The usual maximum-likelihood decoding rule consists in principle of: (i) computing the probability of the hypothesis that a given word has been transmitted, for each codeword; and (ii) choosing the codeword which corresponds to the hypothesis whose probability is the largest. We just propose not to perform step (ii) which makes the result simpler but irreversibly destroys part of the information obtained from step (i).

The practical interest of the proposed redefinition becomes clear when one looks at a coding system where two codes are concatenated i.e., where the result from encoding the data by one of the codes (referred to as outer) is in turn encoded according to the second one (referred to as inner). If decoding the inner code results in a probability distribution instead of a hard decision, it can be used in order to weight the outer decoding, therefore improving it.

As redefined, decoding may be considered as a particular case of the problem of determining a posterior distribution i.e., compatible with a given set of constraints, which is the best one with respect to a given prior probability distribution. According to Shore and Johnson [1], the unique optimum solution to this problem results from applying Kullback's minimum cross-entropy principle [2].

Let $\{p(x_i)\}$ be the prior probability distribution associated with a finite or countable set of elementary events $\{x_i\}$, to be denoted by $p$; we assume that none of the probabilities $p(x_i)$ is zero. Kullback principle consists of choosing, among the distributions $\{q(x_i)\}$ which are compatible with the constraints (to be denoted by $q$), the one which minimizes cross-entropy (denoted by $q^*$). Cross-entropy of $q$ and $p$ is defined as :

$$H(q,p) = \sum_i q(x_i) \log \frac{q(x_i)}{p(x_i)}. \qquad (1)$$

$H(q,p)$ is not symmetric in $p$ and $q$; it is positive and in a certain sense measures the vicinity of $q$ and $p$.

If a block code of length $n$ is to be decoded, the weighting information results from the demodulator evaluating the probability of the hypotheses concerning each of the transmitted symbols. Provided we assume them to be independently affected by the channel errors, a prior probability can be assigned to any $n$-uple by the product of the probability of its symbols, whether this $n$-uple belongs or not to the code. As redefined, decoding consists of recomputing the probability distribution on the $n$-uples in order to take into account the code constraints. The posterior distribution $q^*$ must therefore involve probability zero for all $n$-uples which do not belong to the code. Clearly, the result from interverting $p$ and $q$ in (1) is meaningless, which expresses the decoding irreversibility.

## 2 - REDEFINING MAXIMUM LIKELIHOOD DECODING

Assuming that the exact prior distribution is known, we easily show that Kullback principle results in the set of probabilities which are to be compared in step (ii) of conventional maximum likelihood decoding.

In the binary case (as an example), we consider decoding a particular codeword and assume that the exact prior probability of the symbols is available as the demodulator output. The posterior probability of a given word $\underline{c}_i$ of code C which results from applying Kullback principle is precisely equal to the probability that the transmitted $n$-uple is $\underline{c}_i$, conditioned on the fact it belongs to code C. Therefore applying Kullback principle is equivalent to maximum likelihood decoding in this case.

Kullback principle may also be applied if one assumes that a particular codeword is transmitted (e.g., the all-zero word if the code is assumed to be linear) and considers the probability distribution on the received symbols as resulting from the channel transition probabilities (hence continuous for additive continuous noise if no quantization occurs).

In both cases, the posterior distribution is degenerated in the sense that the contraints result in restricting its support. Hence $q^*$ is proportional to $p$ wherever it is nonzero; the proportionality constant simply renormalizes it and its logarithm is equal to $H(q^*, p)$. It is furthermore possible to restrict the distribution support step by step. We shall make use of this remark in Paragraph 4.

Up to now, we considered only decoding a single code so the posterior distribution is made of $M$ nonzero probabilities, where $M$ is the number of codewords ($M = 2^k$ for a linear binary code having $k$ dimensions, as we shall assume). If several successive decodings are effected, which is precisely the case where it is useful to redefine decoding as we propose it, an extremely fast increase of the alphabet size in terms of the decoding step results. In order to avoid it, but at the expense of strict optimality, we propose to use symbol-by-symbol decoding i.e., we seek posterior probabilities in the form of a product of probabilities separately associated with each information symbol. This approximation yields a great simplification but little impairment and expresses the posterior information in the same form as the prior one, so that successive soft-decision decodings are easily performed. As applied for decoding a linear code, Kullback principle leads to a system of implicit nonlinear equations which can be directly written from its generator matrix. This system can practically be solved by iteration [3].

## 3 - RECEPTION IN THE PRESENCE OF SUBJECTIVITY

Redefining decoding and using Kullback principle has also the advantage that the posterior distribution $q^*$ becomes a continuous function of the prior one, even though the actually available prior distribution may be different from the exact one. Therefore the sensitivity of the decoding result in terms of subjective (approximate or biased) knowledge of $p$ can be studied.

The continuity of $q^*$ with respect to $p$ actually justifies the very practical use of weighted decoding since only approximate weighting information is actually available. Robustness of decoding in terms of the weighting information, which is an experimental fact, finds here a theoretical background.

Since Kullback principle enables computing a posterior distribution whatever the prior one is, its use does not necessarily provide an improvement. We have to determine to what extent it leads to a significant improvement when the available prior distribution $p'$ differs from the exact one $p$.

Let $q'^*$ be the posterior distribution which results from Kullback principle when applied to $p'$. We propose $\mathrm{H}(q'^*, p') + \mathrm{H}(p', p)$ as a criterion such that the smaller it is, the best is decoding. In a sense, it measures the distance between the posterior distribution from decoding and the true prior distribution $p$.

## 4 - ANOTHER CONSEQUENCE OF THE PROPOSED POINT OF VIEW

Another interesting consequence of the proposed viewpoint, both for theoretical and practical reasons, seems to be that decoding, as redefined, does not involve any information loss with respect to the prior distribution. When several codes are associated by product [4] or concatenation [5], successive decodings may be performed with no degradation when their number increases (at least in principle). According to our viewpoint, they are just means for stepwise taking into account partial constraints.

Moreover, very simple codes (hence easily decoded) may be combined. The overall necessary complexity only results from the number of such codes. As an example, we consider Elias-like iterated product codes [4], where parity-check codes are used as components. Provided the initial signal-to-noise ratio is high enough, vanishingly small error probability results with nonvanishing overall rate.

## 5 - CONCLUSION

Redefining weighted decoding seems to be interesting both from a theoretical point of view, especially because it enables understanding the robustness of soft-decision decoding with respect to the weighting information, and a practical one, as improving the outer code decoding in concatenated systems or enabling the design of easily decoded systems where numerous simple codes are combined. Ref. [6] is a more comprehensive discussion of these topics.

## REFERENCES


[1] J.E. SHORE and R.W. JOHNSON, Axiomatic Derivation of the Principle of Maximum Entropy Principle and the Principle of Minimum Cross-entropy, IEEE Trans. Inf. Th., **26** n° 1, jan. 1980, pp. 26-37

[2] S. KULLBACK, Information Theory and Statistics, Wiley, 1959

[3] G. BATTAIL, Décodage Pondéré des Codes Linéaires, IPMU Symposium, Paris, 30 june-4 july 1986

[4] P. ELIAS, Error-free Coding, IRE Trans. **IT**, jan. 1954, pp. 29-37

[5] G.D. FORNEY Jr, Concatenated Codes, MIT Press, 1966

[6] G. BATTAIL, Le Décodage Pondéré en tant que Procédé de Réévaluation d'une Distribution de Probabilité, Annales des Télécommunications, **42**, n° 9-10, sep.-oct. 1987

# A WEIGHTED-OUTPUT SYMBOL-BY-SYMBOL DECODING ALGORITHM OF BINARY CONVOLUTIONAL CODES

---

J.C. BELFIORE
Ecole Nationale Supérieure des Télécommunications
46, rue Barrault, 75634 Paris CEDEX 13 , France.

**Abstract**

**A weighted-output symbol-by-symbol soft-decision decoding algorithm for convolutional codes is described. Its main intended use concerns concatenation schemes. If a convolutional code is used as inner code, it makes soft-decision decoding of the outer code possible, which improves the overall error rate. This algorithm relies on Bayesian estimation. Its comparison with Battail algorithm using cross-entropy minimisation shows that the latter is a special case of the former.**

## 1. Introduction

By concatenation, one can achieve long codes which however can be decoded by two decoders suited to much shorter codes [1]. Considerable savings in decoding complexity thus result, since the overall complexity is the sum of the individual ones and not their product, but at the expense of some sacrifice in performance [2]. If we want to decode this "supercode" optimally, we should weight the outputs of the inner decoder in terms of their likelihood and perform soft-decision decoding of the outer code [3-8]. For instance, Battail proposed a weighted-output version of the Viterbi algorithm [9,10].

This work is intended to a weighted-output symbol-by-symbol decoding algorithm of binary 1/2-rate convolutional codes. Section two introduces the notations used and the basic assumptions made; section three describes the algorithm; section four compares this algorithm with the expression found by Battail [11] by cross-entropy minimization [12]; finally, section five deals with how to initialize the algorithm.

## 2. Notations and Basic Assumptions

We shall study throughout this paper a binary 1/2-rate convolutional code with generator polynomials of the form

$$G^{(i)}(D) = \sum_{n=0}^{L-1} g_n{}^{(i)}(D); \quad (i = 1, 2), \tag{1}$$

L being referred to as the constraint length of the encoder.

An information symbol entering the encoder will be denoted by $u_k$; $t_k{}^{(i)}$, i = 1, 2, will denote the coded symbols delivered at the same instant by it.

White additive Gaussian noise and binary antipodal modulation are assumed (extending the results to other types of modulation is possible). If we denote by $y_i$ the i-th received signal after demodulation, its likelihood function

$$a_i = \ln \frac{Prob\,(u_i = 0 \,|\, y_i)}{Prob\,(u_i = 1 \,|\, y_i)}, \tag{2}$$

is proportional to $y_i$, namely

$$a_i = 4.\sqrt{E_b/N_0}.y_i. \tag{3}$$

The likelihood function $a_i$ is thus a Gaussian variable, whose variance is twice the absolute value of its mean. The transmitted symbols are assumed to be equiprobable.

## 3. The weighted-output algorithm

### 3.1. Expression found by Bayesian estimation

We shall use Bayesian estimation in order to determine the probability that the information symbol $u_k$ is equal to 0.

The transmitted symbols which depend on $u_k$ are

$$t_k{}^{(i)}, \ldots, t_{k+L-1}{}^{(i)} \quad (i = 1, 2), \tag{4}$$

where L is the constraint length of the code. But this set of coded symbols depends itself on the set of information symbols

$$u_{k-L+1}, \ldots, u_{k-1}, u_{k+1}, \ldots, u_{k+L-1}. \tag{5}$$

Therefore, in order to express the probability $Pr\ (u_k = 0)$ in terms of information symbol likelihoods, one should take into account those which belong to the set (5). We shall refer to this set as a "k-superstate" and denote it $S_k{}^r$, where r = 1,..,$2^{2(L-1)}$.

Let $(a_j{}^{(i)})$, i = 1, 2; j = k,..,k+L-1, be the coded symbol likelihoods. Hence, if the superstate $S_k{}^r$ and the received likelihoods $a_j{}^{(i)}$ are assumed to be known, then Bayes's rule results in

$$Prob\ (u_k = 0 \mid S_k^r, (a_j^{(i)})) = \frac{p((a_j^{(i)}) \mid S_k^r, u_k=0)\ .\ Prob\ (u_k = 0)}{p((a_j^{(i)}))}$$

$$= \frac{1 + \tanh\left(1/2\left(\sum_{i=1}^{2} \sum_{m=0}^{L-1} g_m^{(i)} a_{k+m}^{(i)} (-1)^{\sum_{l=0}^{L-1} g_l^{(i)} u_{k+m-l}}\right)\right)}{2} \tag{6}$$

where p(x) is the probability density of the random variable x.

### 3.2. The algorithm

Transmitted symbols are equiprobable. So, $Prob\ (u_k = 0) = 1/2$.

Let $\mu_k(S_k^r)$ denote expression (6). If we consider now every k-superstate $S_k^r$, and sum the products of $\mu_k(S_k^r)$ by the probability of the corresponding k-superstate, we get the probability of deciding $u_k = 0$ given the decoded symbols likelihoods namely,

$$Prob(\ u_k = 0 \mid (a_j^i)) = \sum_{r=1}^{2^{2(L-1)}} \mu_k(S_k^r).Prob(\ S_k^r). \tag{7}$$

Conditioning on $(a_j^i)$ i.e. on the received data, will be omitted in the following as being implicit in any reception problem.

Let $A_k$ be the a posteriori likelihood of symbol $u_k$ i.e., the decoded output concerning this symbol; then $\phi_0(A_k) = \dfrac{1 + tanh\,(A_k/2)}{2}$ is the probability that $u_k$ is equal to 0 and $\phi_1(A_k) = \dfrac{1 - tanh\,(A_k/2)}{2}$ is the probability it is equal to 1. We assume the decoded symbol likelihoods to be separable in the sense that the joint a posteriori probability of several information symbols is the product of their individual probabilities. We have therefore:

$$Prob(\ S_k^r) = \prod_{n=-L+1}^{L-1} \phi_{r_n}(A_{k+n}),$$

where $r_n$ is the n-th digit of the binary representation of integer r i.e. $r = \sum_{n=-L+1}^{L-1} r_n 2^{n+L-1}$. We finally get

$$A_k = \phi_0^{-1}\left(\ \sum_{r=1}^{2^{2(L-1)}} \mu_k(S_k^r).\ \prod_{n=-L+1}^{L-1} \phi_{r_n}(A_{k+n})\ \right), \tag{8}$$

This is an implicit expression of the *a posteriori* symbol likelihoods function of the *a priori* ones.

## 4. Relationship with the result of cross-entropy minimization

We notice that (7) expresses the mean of a random variable so we can write

$$Prob\ (u_k = 0) = \sum_{r=1}^{2^{2(L+1)}} \phi(X_r).Prob\,(S_k{}^r)$$

$$= E\,(\phi(X)),$$

where X is the random variable which takes the value $X_r = \mu_k(S_k{}^r)$ with probability $Prob\,(S_k{}^r)$. Let us assume that the variance of X is small; then

$$E\,(\phi(X)) \sim \phi(E\,(X))$$

so that (8) approximately reduces to

$$A_k = \sum_{i=1}^{2} \sum_{n=0}^{L-1} g_n{}^{(i)} a_n{}^{(i)} \frac{\prod_{l=0}^{L-1} \tanh(A_{k-n+l}/2)}{\tanh(A_k/2)}, \tag{9}$$

which is the expression found by Battail using the minimum cross-entropy principle 11.

We notice that, for a non-systematic code, $A_k = 0$ for $k \in N$, is a solution of (9), for every set of $a_n{}^{(i)}$. The simulations performed for certain non-systematic codes showed that $A_k$ converged to 0. This does not occur with (8) since 0 is not a solution of it.

## 5. Initialisation

In (8) or in (9), $A_k$ depends on $A_{k+1},...,A_{k+L-1}$; we thus need an initial estimate of these symbols. In the hard decision sense, a good one results from inverting the generator matrix G of the code, since a pseudo-inverse $G^{-1}$ of it exists, such that

$$G\ .\ G^{-1} = D^t \quad (t \in N\ )\ . \tag{10}$$

We can apply this pseudo-inverse to the received likelihoods $a_n{}^{(i)}$ by replacing modulo 2 addition of binary symbols by the operation on the corresponding likelihoods (real numbers) defined as

$$a\ \Delta\ b = Log\ \frac{1 + \tanh(a/2)\tanh(b/2)}{1 - \tanh(a/2)\tanh(b/2)}\ , \tag{11}$$

which merely results from the identity

$$Prob\ (u + v = 0) = Prob\ (u = 0).Prob\ (v = 0) + Prob\ (u = 1).Prob\ (v = 1)\ .$$

## 6. Conclusion

We described a new symbol-by-symbol decoding algorithm which delivers weighted decisions usable by a subsequent decoder. It extends to convolutional codes, in a sense, the algorithm found by Battail for block codes by minimizing cross-entropy between a priori and a posteriori probability distributions.

***Acknowledgments***

I wish to thank Prof. G. Battail for his suggestions and for his helpful comments during the preparation of this paper.

******* *REFERENCES* *******

1. G. FORNEY, Concatenated codes. Cambridge, Mass.: M.I.T. Press, 1966.

2. G.W. ZEOLI, "Coupled decoding of block-convolutional concatenated codes," Ph.D. dissertation, Dep. Elec. Eng., Univ. California, Los Angeles, 1971.

3. J.K. WOLF, "Efficient maximum likelihood decoding of linear block codes using a trellis," I.E.E.E. Trans. on I.T., Jan. 1978.

4. G. BATTAIL, "Décodage pondéré optimal des codes linéaires en blocs I," Annales des Télécommunications, Nov.-Déc. 1983.

5. C.R.P. HARTMANN and L.D. RUDOLPH, " An optimum Symbol-by-Symbol decoding rule for linear codes," I.E.E.E. Trans. on Inf. Theory, pp. 514-517, 1976.

6. G. BATTAIL et al., " Replication Decoding, I.E.E.E. Trans. on Inf. Theory, pp. 332-345, 1979.

7. L.D. RUDOLPH et al., " Algebraic Analog Decoding of linear binary codes," I.E.E.E. Trans. on Inf. Theory, pp. 430-440, 1979.

8. G.D. FORNEY, "Generalized minimum distance," I.E.E.E. Trans. on Information Theory

9. G. BATTAIL, " Weighting the symbols decoded by the Viterbi Algorithm," I.E.E.E. International Symposium on Information Theory, Oct. 1986.

10. G. BATTAIL, "Pondération des symboles décodés par l'algorithme de Viterbi," Annales des Télécommunications, to be published.

11. G. BATTAIL, " Weighted decoding as a means for reestimating a probability distribution, " this issue.

12. J.E. SHORE and R.W. JOHNSON, " Axiomatic derivation of the principle of maximum entropy and the principle of minimum cross-entropy," I.E.E.E. Trans. on Inf. Theory pp.26-37, 1980.

# SEARCH FOR SEQUENCES WITH ZERO AUTOCORRELATION

by Roger ALEXIS

TRT, 5 avenue Réaumur - 92350 Plessis-Robinson

__Abstract__ : It exists sequences of length N of constant amplitude and zero autocorrelation over 2n symbols, provided that each sequence be preceeded by n symbols and followed by n other symbols. Methods for obtaining these N + 2n symbols are proposed and examples are given.

__Résumé__ : Il existe des séquences de longueur N, à amplitude constante et à fonction d'autocorrélation nulle sur 2n symboles, pourvu que chaque séquence soit précédée par n symboles et suivie par n autres symboles. On propose des méthodes pour obtenir ces N + 2n symboles et on donne des exemples.

## 1 - INTRODUCTION

The search for sequences with good autocorrelation properties gave rise to many works. The reason is the application of such sequences to synchronization of communication systems, to the estimate of the impulse response of a transmission channel (wire or radio) in order to implement some types of equalizers or adaptive detectors, at last to the compression of radar pulses.

Sequences with constant amplitude and zero autocorrelation ("CA-ZAC") are scarce[1] :

- if the aperiodic autocorrelation is considered, the problem has obviously no solution, because a sequence of length N leads to a non zero autocorrelation function for odd shifts ;

- if the periodic autocorrelation is considered, only polyphase sequences exist. Binary sequences can be found with length 4, quaternary with length 8,...., $m^{K+1}$ ary with length $m^{2K+1}$ : the choice is too much limited.

For evaluating the response of a noisy channel, a maximum length pseudo-random sequence is often used. It has been shown [1] that in the presence of noise, the variance of the error was then superior by 3 dB to the variance obtained with a "CAZAC" sequence.

Another method is to use two GOLAY complementary sequences, doubling, by addition, the autocorrelation peak and cancelling the side-lobes. This solution requires to switch off the transmission of signal before, between and after the sequences. These cuts can be unacceptable ; dead times are 50% higher than with the proposed solution ; at last the spreading of the signal in time makes this solution more sensitive to Doppler effect and to drifts of pilots.

The proposed solution consits in searching for sequences having a central word of N bits, preceeded by $n_1$ bits and followed by $n_2$ bits. These $n_1+N+n_2$ bits are chosen in such a way that the aperiodic correlation with the N bits of the central part be equal to :

$$\underbrace{X,\ldots,X}_{N-1},\ \underbrace{\emptyset,\emptyset,\ldots\emptyset}_{n_1},N,\ \underbrace{\emptyset,\ldots,\emptyset,\emptyset}_{n_2},\ \underbrace{X,\ldots,X}_{N-1}$$

the N-1 values marked X on both sides having any values. In order to obtain $n_1+n_2$ zeros, N must be even.

The field of search is reduced if we are limited to the case $n_1 = n_2 = n$ . It can be shown that only this case in of interest when we are looking for the impulse response of a dispersive channel.

In order to be well separated, the channel impulse response must not spread over more than (n+1) consecutive symbols among the (2n+1) symbols formed by the central part of the correlation. The aim of the frame synchronization is then to define the place where this response is expected. It can be shown that this place corresponds to a maximum of received energy integrated on n consecutive symbols.

## 2 - METHOD N$^r$1

Sequences as described above are searched in the binary and quaternary cases, using a "brute force attack", which consists to try the $2^N$ (binary) or $4^N$ (quaternary) possible sequences. Starting from the peak of the autocorrelation, for every shift of one unit towards the right or the left, one degree of freedom is at disposal, consisting in the choice of one of the 2n symbols to maintain the autocorrelation equal to zero.

It is clear that the computation load increases exponentially with N, which limits the possibities of this method.

## 3 - METHOD N$^r$2

An algorithm bringing down the number of trials, in the biphase case, from $2^N$ to $\binom{N}{N/2}$ has been used. It is based on the remark that, in order to get a zero of the autocorrelation for a shift equal to one, it is necessary to have $\frac{N}{2}$ transitions from 1 to 0 or from 0 to 1 in the word of length N.

The algorithm used, allows the partition of a set of N bits into a given number P of subsets. The partition is defined by the ranks $i_1$, $i_2$, ... $i_{P-1}$ of the border delimiting the subsets. All the possible values are given to indexes i, with the condition

$$0 < i_1 < i_2 < \ldots < i_{P-1} < P$$

This is obtained with P-1 imbricated iteration loops.

## 4 - RESULTS OBTAINED

### 4.1 - Binary case

All the sequences have been obtained up to N = 20 with method N$^r$1 and for N = 32 with method N$^r$2 ; their number is indicated in Table 1. These numbers are multiple of two because to each sequence corresponds its inverse.

Tables 2 and 3 give, as examples, the "generating sequences" for N = 18 and N = 32. The sequences are listed in increasing values, read from right to left. For an easier notation, bits are labeled 0 and 1 instead of +1 and - 1.

The whole set of sequences can be deduced from the "generating sequences" by right-left symmetry, inversion (1 becomes 0 and 0 becomes 1) and if N=4L (L integer) by circular permutation.

Thus, each "generating sequence" produces a maximum number of 4 sequences and, if N=4L, of 4N sequences. But, when the "generating sequence" presents a particular structure (symmetry of its inverse or periodic with period $\frac{N}{2}$ or $\frac{N}{3}$) the number of "generating sequences" is reduced. This particular structure does not exist for $16 \leq N \leq 20$. It is conjectured that this is also true for N > 20.

Some other interesting properties can be observed :

a) For $8 \leq N \leq 20$, sequences exist up to $n_{max} = \frac{N}{2} - 1$.

b) The values of the 2n bits located at the right and at the left of the word of length N can be deduced from it easily :

- if N=4L (L integer), the structure is the following :

$$\overbrace{a_{N-i}, \ldots, a_{N-1}, a_N}^{n}\ ,\ \overbrace{a_1, a_2, \ldots, a_N}^{N}\ ,\ \overbrace{a_1,\ a_2, \ldots, a_i}^{n} \quad \text{with } i \leq \frac{N}{2} - 1$$

so that the searched aperiodic structure is in fact a periodic one (example : N = 32).

- if $N \neq 4L$ (L integer) the structure becomes :

$$\overbrace{a_{N-i},\ldots,a_{N-1},\overline{a_N}}^{n}\,,\;\overbrace{a_1,a_2,\ldots,a_N}^{N}\,,\;\overbrace{\overline{a_1},\,a_2,\ldots,a_i}^{n} \quad \text{with } i \leq \frac{N}{2}-1$$

expression where $\overline{a_N}$ and $\overline{a_1}$ represent respectively the inverse of $a_N$ and $a_1$. The structure is no longer periodic (example N = 18).

## 4.2 - Quaternary case

The sequences have been obtained for N = 8, 10 and 12 with method N$^r$1 ; their number is indicated in Table 4. These numbers are multiple of eight because to each sequence corresponds 3 other sequences deduced by rotation of $\frac{\pi}{2}$, $\pi$ and $\frac{3\pi}{2}$, and to the 4 sequences obtained in this manner correspond those deduced by a change of the direction of rotation of the phase.

Table 5 gives, as example, the "generating sequences" for N = 12. The sequences are listed in increasing values, read from right to left. The 4 phases are labeled 0, 2, 1 and 3 instead of 1, i, - 1, - i respectively.

A "generating sequence" beginning at the right with a "1", the change of the direction of rotation of the phase is obtained by exchanging "2" and "3" everywhere. A rotation of $\frac{\pi}{2}$ is obtained by adding "1" modulo 4.

For N = 8, sequences exist up to $n_{max} = N - 1$ : these are the polyphase codewords.

For N = 10, $n_{max} = \frac{N}{2} - 1$, as for the binary case.

For N = 12, $n_{max} = \frac{N}{2}$ : the number of zeros is larger in quaternary case than in binary one.

Finally, the periodical structure of the sequence exists for N = 8 but does not exist for N = 10 and 12, contrary to the binary case.

## 5 - METHOD N°3

### 5.1 - General

This method, valid only for biphase words of length N=4L is inspired from references [2] and [3], where is described a method trying to perform the synthesis of a word having a given aperiodic autocorrelation function. We just give a summary of the method.

First formulae analoguous to those of [3], have been established in the case of a periodic autocorrelation function (case of N=4L).

A word is represented by the list of the lengths of runs of consecutive +1 or -1, i.e. :

$$[R_i] = \{r_1, r_2, \ldots, r_i\} .$$

This notation presents the advantage not to distinguish a word and its inverse.

It is possible to establish conditions for having a zero of the autocorrelation successively at a distance of 1, 2 ... of the peak. When the distance increases, conditions become more and more complex ; so we limited ourselves to the four following conditions :

a) first zero : $[R_i]$ must have $\frac{N}{2}$ terms.

b) second zero : There are $\frac{N}{4}$ runs of length 1 in the word (of length N).

c) third zero : We slide a window of length 2 along the word. There are as many windows with one run of length 2 than windows with two runs of length 1.

d) fourth zero : There are as many windows of length 3 with one run of length 3 or three runs of length 1, than windows (of length 3) with one run of length 2 and one run of length 1.

5.2 - Implementation

The method requires 3 steps, shown on the example N = 32.

a) Step 1 :

We apply the same algorithm as in method 2 to establish the set $\mathcal{P}_1$ of partitions of the $\frac{N}{4} = 8$ runs of length 1 into P packets, P varying from 1 to $\frac{N}{4}$.

Then we eliminate partitions which can be deduced by circular permutation or right-left symmetry. Let $\pi_P$ be the remaining number of partitions into P packets.

For example the length of remaining partitions for N = 32 is :

P = 1 : 8 ($\pi_1 = 1$)
P = 2 : 17,26,35,44 ($\pi_2 = 4$)
P = 3 : 116,125;134,224,233 ($\pi_3 = 5$)
.
.
.
P = 8 : 11111111 ($\pi_8 = 1$)

(and $\pi_4 = 8$ ; $\pi_5 = 5$ ; $\pi_6 = 4$ ; $\pi_7 = 1$)

We can also determine for each partition the number $\omega_{3,3}$ of windows of length 3 with three runs of length 1.

b) Step 2 :

We apply again the same algorithm to establish the set $\mathcal{P}_2$ of partitions of the ramaining $\frac{3N}{4} = 24$ bits into $\frac{N}{4} = 8$ runs of length $\geq 2$. In practice, we substract one to each run and apply the algorithm to $\frac{N}{2} = 16$ bits only ; then we add one to each of the $\frac{N}{4} = 8$ runs obtained.

Let P' the number of runs of length > 2 and $\pi'_{P'}$ the number of partitions for every value of P'. For example, the partitions for N = 32 are :

<u>P' = 1</u> : 2222222$\overline{10}$ ...etc.($\pi'_1$ = 8, shifting the $\overline{10}$ to all possible places)

<u>P' = 2</u> : 22222239, 22222248, 22222257, 22222266,... etc.

($\pi'_2 = \frac{8}{6!}.(3 + \frac{1}{2!}) = 196$, shifting the two numbers differing from 2 to all possible places).

(and $\pi'_3 = 1176$ ; $\pi'_4 = 2450$ ; $\pi'_5 = 1960$ ; $\pi'_6 = 588$ ; $\pi'_7 = 56$ ; $\pi'_8 = 1$).

We can also determine for each partition the number $\omega_{3,1}$ of windows of length 3 with one run of length 3. (it is simply the number of runs of length 3).

c) <u>Step 3</u> :

We apply a third time the same algorithm to cut each set of $\frac{N}{4} = 8$ runs obtained is step 2 into P packets. We remark that this computation is identical to the one performed in step 1 (without the elimination which followed).

We determine, for each partition the number $\omega_{3,2}$ of windows of length 3 with one run of length 1 and one run of length 2 : it is simply the number of "2" located on the edges of the P packets (if a packet consists only of one run of length 2, it is counted twice).

Finally, for a given value of P, we interleave the P packets obtained in step 3 with the P packets obtained in step 1, only if conditions c) and d) of paragraph 5.1 above are satisfied. The condition c) limits the associations to the case P = P'.

The condition d) can be written $\eta_{3,2} = \omega_{3,1} + \omega_{3,3}$ and must be checked for every class of sequences.

For example, for P = P' = 2, the partition 22222239 of runs of length ≥ 2 (step 2) will be cut and interleaved with the packets of 3 and 5 runs of length 1 (step 1) to give the following sequences :

$$\left.\begin{array}{l} 1\ 1\ 1\ 2\ 1\ 1\ 1\ 1\ 1\ 2\ 2\ 2\ 2\ 2\ 3\ 9 \\ 1\ 1\ 1\ 2\ 2\ 1\ 1\ 1\ 1\ 1\ 2\ 2\ 2\ 2\ 3\ 9 \\ \quad . \\ \quad . \\ \quad . \\ 1\ 1\ 1\ 2\ 2\ 2\ 2\ 2\ 2\ 3\ 1\ 1\ 1\ 1\ 1\ 9 \end{array}\right\} \text{ or } \binom{7}{1} = 7 \text{ sequences}$$

Each partition obtained in step 2 give rise, for a given P, to $\binom{N/4-1}{P-1}$ possible cuts.

Thus, the total number of sequences to consider is :

$$\sum_{P=1}^{P=N/4} \pi_{P'} \cdot \pi'_{P'} \cdot \binom{N/4-1}{P-1}$$

instead of $2^N$ with the method N° 1. For N = 32, it corresponds to a gain of about 3500 on the number of sequences to consider. At present, this method has been validated for N = 32.

<u>Remark</u> : The computation time can be still reduced, remarking that some associations don't need to be considered. For example :

if $2P < \frac{N}{4} - 2$, there is the condition $\omega_{3,3} + \omega_{3,1} \leq 2P$

if $2P = \frac{N}{4}$ , there is the conditions $\omega_{3,3} > 0$ and $\omega_{3,1} > 0$

if $2P > \frac{N}{4} + 2$, there is the condition $\omega_{3,3} + \omega_{3,1} \leq \frac{N}{4} - P$ .

These conditions allow to eliminate entire classes of sequences, without having to consider them individually.

## 5 - FURTHER POSSIBILITIES

Above results have demonstrated that there was a simple relationship between the N symbols and the 2n symbols of a sequence. This suggests that there exists probably some hidden algebric properties to discover, allowing an easier determination of such sequences.

## REFERENCES


[1] A. MILEWSKI "Periodic sequences with optimal properties for channel estimation and fast start up equalization". IBM J. Res. Develop. sept. 1983 pp 426 - 431.

[2] R. POLGE et al "A new technique for the design of binary sequences with specified correlation" - IEEE Southeastcon 1981, Huntsville Alabama, pp 164 - 169.

[3] R. POLGE "A general solution for the synthesis of binary sequences with desired correlation sequences" - AGARD conference 381 - pp. 25.1 - 25.9. Multifunction radar for airborne applications - Toulouse, Oct 85.

| N \ n | 1 | 2 | 3 | 4 | 5 | 6 | 7 | 8 | 9 | - - - | 12 | 13 | 14 |
|---|---|---|---|---|---|---|---|---|---|---|---|---|---|
| 8 | 70x2 | 36x2 | 20x2 | 0 | | | | | | | | | |
| 10 | 252x2 | 60x2 | 12x2 | 4x2 | 0 | | | | | | | | |
| 12 | | 400x2 | 184x2 | 120x2 | 72x2 | 0 | | | | | | | |
| 14 | | | 132x2 | 20x2 | 4x2 | 0 | | | | | | | |
| 16 | | | | 768x2 | 256x2 | 128x2 | 64x2 | 0 | | | | | |
| 18 | | | | | 36x2 | 12x2 | 4x2 | 0 | | | | | |
| 20 | | | | | | 100x2 | 440x2 | 200x2 | 40x2 | 0 | | | |
| . | | | | | | | | | | | | | |
| . | | | | | | | | | | | | | |
| . | | | | | | | | | | | | | |
| . | | | | | | | | | | | | | |
| . | | | | | | | | | | | | | |
| 32 | | | | | | | | | | | 640x2 | 512x2 | 0 |

N.B. : When N is a multiple of 4, sequences are deduced by circular permutation.

TABLE 1

Number of sequence (binary)

| Number of words | Generating sequences | | |
|---|---|---|---|
| 4 | 0 1 0 0 1 | 1 1 1 0 1 0 1 1 1 0 0 1 0 0 1 0 0 0 | 0 1 1 0 1 |
| 4 | 0 1 0 0 1 | 0 0 0 1 0 1 0 0 1 1 1 1 0 0 1 0 0 0 | 1 0 0 1 0 |
| 4 | 0 0 0 1 0 1 | 0 0 1 0 0 0 1 0 1 1 1 0 0 0 0 1 0 0 | 1 0 1 0 0 0 |
| 4 | 0 0 1 0 1 | 1 1 0 1 1 0 1 0 1 1 1 0 0 0 0 1 0 0 | 0 1 0 1 1 |
| 4 | 0 0 1 0 1 | 0 0 1 0 0 0 1 1 1 1 0 1 0 0 0 1 0 0 | 1 0 1 0 0 |
| 4 | 0 0 1 0 1 | 1 1 0 1 1 0 1 0 1 1 1 1 0 0 0 1 0 0 | 0 1 0 1 1 |
| 4 | 1 1 0 1 0 1 | 1 1 0 1 0 0 0 0 1 0 0 0 1 1 0 1 0 0 | 0 1 0 1 0 0 |
| 4 | 0 1 1 0 1 | 0 0 1 1 0 1 0 1 0 0 0 0 0 0 1 1 0 0 | 1 0 1 1 0 |
| 4 | 0 1 0 1 1 0 1 | 1 1 0 0 1 0 1 0 0 0 0 0 1 0 1 1 0 0 | 0 1 0 0 1 |
| 4 | 0 0 0 1 1 | 0 1 0 0 0 1 1 1 0 0 1 0 0 0 0 0 1 0 | 1 1 0 0 0 |
| 4 | 0 0 0 1 1 | 1 0 1 1 1 0 0 1 0 1 1 0 0 0 0 0 1 0 | 0 0 1 1 1 |
| 4 | 0 0 1 1 1 | 1 0 0 1 1 0 1 0 1 0 0 0 0 0 0 1 1 0 | 0 0 0 1 1 |
| 4 | 0 0 0 0 1 1 1 | 0 1 1 0 0 0 0 1 0 1 0 0 0 0 0 1 1 0 | 1 1 1 0 0 0 0 |
| 4 | 0 1 1 1 1 | 0 1 1 1 0 1 0 0 1 0 0 0 0 0 1 1 1 0 | 1 1 1 1 0 |
| 4 | 0 1 1 1 1 | 0 1 1 1 0 1 1 0 1 0 0 0 0 0 1 1 1 0 | 1 1 1 1 0 |
| 4 | 1 0 1 1 1 1 | 1 0 0 0 1 0 0 0 0 1 0 0 1 0 1 1 1 0 | 0 0 0 0 1 0 |
| 4 | 0 1 1 1 1 | 1 0 0 0 1 0 0 0 0 1 0 1 1 0 1 1 1 0 | 0 0 0 0 1 |
| 4 | 0 1 1 1 1 1 | 0 1 1 1 1 0 1 0 0 0 1 0 0 1 1 1 1 0 | 1 1 1 1 1 0 |

-------
72 for n = 5
24 for n = 6
8 for n = 7

**TABLE 2**

**Words of length N = 18 (n = 5, 6, 7)**

| Number of words | Generating sequences | | |
|---|---|---|---|
| 128 | 0100110000000 | 11010110011111001010100110000000 | 1101011001111 |
| 128 | 011001000000 | 10111010110001101000011001000000 | 101110101100 |
| 128 | 1010011000000 | 11101011001111101101010011000000 | 1110101100111 |
| 128 | 010000100000 | 11011000010110011101010000100000 | 110110000101 |
| 128 | 0111000100000 | 11010000110111011010111000100000 | 1101000011011 |
| 128 | 1000100100000 | 11101110010111100101000100100000 | 1110111001011 |
| 128 | 0111100100000 | 11010001110111011010111100100000 | 1101000111011 |
| 128 | 1000110100000 | 11101110110111100101000110100000 | 1110111011011 |
| 128 | 1101000010000 | 11000101111011001011101000010000 | 1100010111101 |
| 128 | 0011100010000 | 10111011011011110100011100010000 | 1011101101101 |

-----
1280 for n = 12
1024 for n = 13

TABLE 3

Words of length N = 32 (n = 12, 13)

| N \ n | 1 | 2 | 3 | 4 | 5 | 6 | 7 | 8 |
|---|---|---|---|---|---|---|---|---|
| 4 | | 8x8 | 8x8 | 0 | | | | |
| 6 | | 42x8 | 6x8 | 2x8 | 0 | | | |
| 8 | | 536x8 | 160x8 | 16x8 | 16x8 | 16x8 | 16x8 | 0 |
| 10 | | | 632x8 | 80x8 | 0 | 0 | | |
| 12 | | | | 1056x8 | 492x8 | 4x8 | 0 | |

TABLE 4

Number of sequences (quaternary)

| Number of words | Generating sequences | | |
|---|---|---|---|
| 8 | 0 3 1 1 1 1 | 2 3 3 2 1 2 1 2 1 1 0 0 | 3 2 3 2 0 3 |
| 8 | 1 3 0 1 0 1 | 3 3 2 2 0 2 0 2 0 1 1 0 | 2 2 2 2 1 3 |
| 8 | 2 3 2 0 3 1 | 0 2 0 2 2 3 3 2 2 0 2 0 | 1 3 0 2 3 2 |
| 8 | 2 2 2 0 3 1 | 1 3 1 3 2 2 3 3 2 0 2 0 | 0 2 1 3 3 3 |
| ---- | | | |
| 32 | | | |

TABLE 5

Words of length N = 12 (n = 6)

# Adapted Codes For Communication Through Multipath Channel
# Codages Adaptés à la Communication avec Trajets Multiples

**G. Hakizimana, G. Jourdain, G. Loubet**

CEPHAG, UA 346, INPG/IEG, BP 46, F-38402 ST Martin d'Hères Cedex

**Abstract :** *Two systems for communication through multipath channels are presented and compared here. They are based on coding with binary sequences. The first one uses a* ***pseudo-orthogonal coset code [1]*** *and the second one the* ***Yates-Holgate sequences [8,5].*** *Both are briefly described and their communication properties are recalled. An underwater experiment has been conducted with these codes. The obtained performances are given in terms of error and transmission rates. Two reception schemes are compared.*
**Key words:***Multipath channels, communication, binary sequences, correlation.*

**Résumé :** *Deux systèmes de communication à travers des canaux à trajets multiples sont étudiés et comparés. Ils sont basés sur le codage à séquence binaires. Le premier utilise un* ***code pseudo-orthogonal coset [1]*** *et le second les* ***séquences de Yates et Holgate [8,5].*** *Nous rappelons leurs propriétés. Nous avons réalisé une expérience de communication sous-marine utilisant ces deux types de codage. Les résultats obtenus sont présentés en termes de taux d'erreur et de débit. Nous comparons deux types de récepteurs.*
**Mots-clés:***Trajets multiples, communication, séquences binaires, corrélation.*

## 1) INTRODUCTION

Multipath phenomenon occurs in many natural communication channels and especially in the underwater channel. It leads to fading and intersymbol interference which seriously disturb the communication. Many techniques have been studied in order to improve the transmission quality : mainly, array treatment, use of multiples frequencies, equalization and coding spread-spectrum techniques. The latter has shown to be much robust and more efficient, especially when the signal-to-noise ratio (SNR) is small. We have used it henceforth.

In this paper, we present and compare two communication systems based on binary spread-spectrum sequences. The choice of sequence family depends of course on the performance to be achieved, but also on several parameters : some of them are connected with the channel and other to the complexity of the receiver scheme.

In the first part, we recall the channel model and we present briefly the selected codes. The first is the ***Gold or pseudo-orthogonal code*** [6] described by Chavand [1]. The second was developed by Yates and Holgate [8]. and generalized by Potter [5]. The first one have been tested in ionospheric communication and we have compared them in underwater case to other similar codes [2]. On the other hand, the ***Yates-Holgate sequences*** have not been used in natural channel, to our knowledge.

In the second part, an underwater experiment using these sequences is presented. The obtained performances are presented in term of error rates observed and transmission rate achieved. Two reception schemes are compared.

## 2) CHANNEL MODEL AND COMMUNICATION SCHEME

**2.1) Channel model** : Let s(t) be the transmitted signal, r(t) the received one, and b(t) an additive noise. The assumed base band multipath channel model is described by :

$$r(t) = \sum_{i=0}^{L-1} a_i \, s(t-t_i) \, e^{j\theta i} + b(t) \qquad (1)$$

So, the channel consists of L paths of strengths $a_i$, delays $t_i$ and phases $\theta_i$. b(t) is assumed to be white, gaussian and zero-mean. In a general case, the quantities $a_i$, $t_i$ and $\theta_i$ can be known, constant, time varying, or random.

**2.2) Communication schemes :**

**Emission** : to encode the data, a N length codeword is associated with k information bits. The codeword is a binary sequence (+1,-1) belonging to a set (code) of $M = 2^k$ sequences, all of the same length. These sequences lead to signals with the same bandwidth, energy and duration T.

**Reception** : the problem in a M-ary communication is to distinguish one signal among the M allowed ones. In order to do this, this system uses a bank of M correlators as receiver. This is the well-known optimal receiver for a single-path channel with additive white noise. The largest output indicates the received codeword (**Fig. 1**).

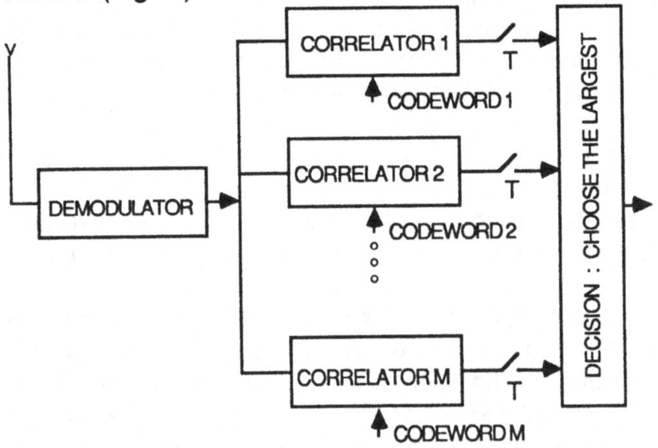


**Figure 1 : Optimal Receiver for Single-Path Channel**

In a multipath situation, this receiver is very sensitive to partial correlations due to delayed signals. Thus, keeping this receiver involves sets of sequences with low auto-correlation sidelobes and also small peak cross-correlation magnitude for any delay. The pseudo-orthogonal codes satisfy these criteria.

The proposed scheme using Yates-Holgate sequences **[8]** is quite different : the set of sequences are all derived from a single m-sequence, called the 'parent' m-sequence.

When any sequence is cross-correlated with the 'parent', the resulting cross-correlation function has a unique shape which simply and uniquely identifies the sequence. This system leads to a very simplified receiver which consists only in one correlator (**Fig. 2**).

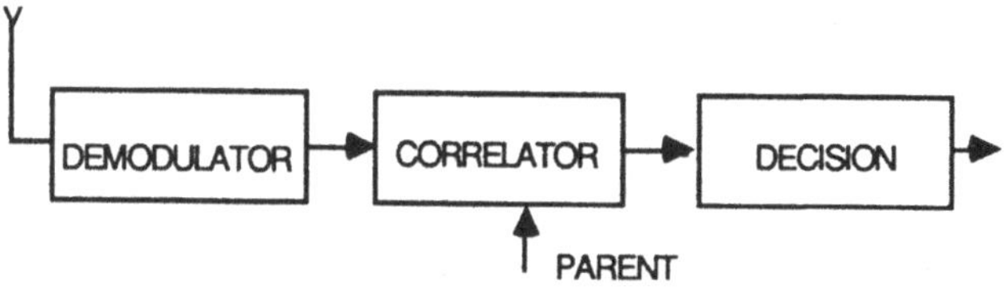


**Figure 2 : Receiver for Yates-Holgate Sequences**

## 3) PROPERTIES OF SEQUENCE FAMILIES

**3.1) Periodic and aperiodic correlation, efficiency** : The receiver in Fig. 1 uses aperiodic correlation. But, in order to select optimal codes in the sense defined above, the first step is the use of periodic correlation because it is analytically computable. The second step will consist in numerical selection of the codes that keep good aperiodic properties among those which had good periodic ones. Let S be a code or set of M sequences of length or period N ; **X** and **Y** two sequences in S. Let $\mathbf{X}=(X_0,\ldots, X_{N-1})$ and $\mathbf{Y}=(Y_0,\ldots, Y_{N-1})$, the periodic correlation between **X** and **Y**, $\Gamma^p$ and the aperiodic one $\Gamma$ are defined as

$$\Gamma^p_{XY}(m)=\frac{1}{N}\sum_{r=0}^{N-1} X_r Y_{r+m} \quad ; \quad \Gamma_{XY}(m)=\frac{1}{N}\sum_{r=0}^{N-1-m} X_r Y_{r+m} \qquad (2)$$

$$\Gamma^p_a = \mathrm{Max}\ \{\ |\ \Gamma^p_{XX}(m)\ |,\ 1 \le m \le N\text{-}1\ ;\ \ \mathbf{X}\in S\ \} \qquad (3)$$

$$\Gamma^p_i = \mathrm{Max}\ \{\ |\ \Gamma^p_{XY}(m)\ |,\ 0 \le m \le N\text{-}1\ ;\ \ \mathbf{X},\mathbf{Y}\in S\ ,\ \mathbf{X} \ne \mathbf{Y}\ \} \qquad (4)$$

$$\Gamma^p_{Max} = \mathrm{Max}\ \{\ \Gamma^p_a\ ,\ \Gamma^p_i\ \} \qquad (5)$$

The subscript r+m in (2) is taken modulo N. In the aperiodic case, the parameters are $\Gamma_a, \Gamma_i, \Gamma_{Max}$ and are defined in the same way, replacing in (3), (4) and (5) $\Gamma^p$ by $\Gamma$.

Let $\varepsilon = \text{Max} \{ \, | \Gamma_{XY}(0) | \, , \, \mathbf{X},\mathbf{Y} \in S \, , \, \mathbf{X} \neq \mathbf{Y} \}$ (6)

Finally, the last important parameter is the **code rate** or **efficiency** :

$$\alpha_R = k/N \qquad \text{where } k = \log_2 M \tag{7}$$

With the Fig.1 receiver, a good code will provide high $\alpha_R$ and small $\Gamma_{Max}$ and $\varepsilon$.

**3.2) Description of pseudo-orthogonal coset codes** : These code are described by F. Chavand **[1]** and are constructed as Gold codes **[6]**. They consist of $2^k$ sequences of length $2^k$-1. It is shown that for these codes

$$N \Gamma^p_{Max} = 2^{[(k+2)/2]} \pm 1$$

$$\alpha_R = k / 2^k - 1 \tag{8}$$

where [u] denotes the interger part of the number u.
It is called pseudo-orthogonal because

$$| N\Gamma_{XY}(0) | = 1 \, , \quad \forall \quad \mathbf{X} \neq \mathbf{Y} \tag{9}$$

The selected code is a Gold one. It is the so-called "V45 C75 Rot0" **[1]** which is a 31 length coset code. Its parameters are

$$\Gamma^p_{max} = 9/31, \; \Gamma_{max} = 11/31, \; \alpha_R = 5/31 \tag{10}$$

The set size is 32. This code achieves the best trade-off between $\alpha_R$ and $\Gamma_{max}$ in comparison with other lengths. Besides, it is optimal with respect to the periodic correlation bounds given by Welch and Sidel' Nikov **[6]**.

**3.3) The Yates-Holgate Code** : Let $\mathbf{M} = (M_r)$ denotes a N length m-sequence whose elements $M_r$ are taken from {-1,+1}.

**a)** The shift-and-add property of m-sequences can be expressed as :
$\forall$ d $\neq$ 0, there exists d* $\neq$ 0, d such as

$$M_r \cdot M_{r-d} = M_{r-d^*} \tag{11}$$

d and d* are modulo N defined.

Let h(x) be the recurrence polynomial of the m-sequence **M**. A way to obtain d* from d and h(x) is given in **[4]**. It is based on the increasing power division of the polynomial $(1 + x^d)$ by h(x)

$$\frac{1+x^d}{h(x)} = q(x) + \frac{x^{d^*}}{h(x)} \tag{12}$$

The division is performed until a one-term remainder is obtained. Its exponent is d*. We note that the couples (d,d*) can be directly numerically obtained.

**b) Construction of Y-H families :** with one m-sequence **M** , Potter **[5]** constructs many different sets of Yates-Holgate (Y-H) sequences. Let **Z**(i) denotes a Y-H sequence belonging to the $i^{th}$ set.

$$Z_r(i) = (1-2^{1-i})\, M_r - 2^{1-i} \sum_{k=1}^{2^i-1} M_{r-\phi(i,k)} \tag{13}$$

where : $\phi(i,k) \neq 0$ ; $\phi(i,k) \neq \phi(i,j)$ ; $\phi(i,k) \neq \phi*(i,j)$ for $j \neq k$ (14)
$\phi$ and $\phi*$ are related through (11).

It is shown in **[5,8]** that **Z**(i) is binary and balanced. In order to construct a sequence belonging to the $i^{th}$ family, $2^i$-1 phases $\phi(i,k)$ satisfying (14) are necessary. There exists a relationship between these phases :

$$\phi(k \oplus j) = [\ \phi(k) + [\ \phi(j) - \phi(k)]^*\ ]^* \tag{15}$$

where k and j are base-2 represented (i.e. $3 \oplus 1 = 2$).This relation allows the sequence construction. It is shown **[5]** that i phases are sufficient to construct the sequence. The $2^i$-1-i other phases are obtained from the first ones owing to the relation (15). The correlation expressions are obtained from (13) and (2)

$$\Gamma^p_{\mathbf{Z(i)\,M}}(m) = (1-2^{1-i})\,\Gamma^p_{\mathbf{M}}(m) - 2^{1-i} \sum_{k=1}^{2^i-1} \Gamma^p_{\mathbf{M}}(m-\Phi(i,k)) \tag{16}$$

where $\Phi(i,k) = N - \phi(i,k)$

By recalling that : $\Gamma^p_{\mathbf{M}}(0) = 1$ and $\Gamma^p_{\mathbf{M}}(m) = -1/N$ , $\forall$ $m \neq 0$ , one obtains :

$$\Gamma^p_{\mathbf{Z(i)\,M}}(0) = [1 + (1-2^{1-i})(N+1)]/N$$

$$\Gamma^p_{\mathbf{Z(i)\,M}}(\Phi(i,k)) = [1 - 2^{1-i}(N+1)]/N \quad ; \ k \in \{1,2,\ldots,2^i-1\} \tag{17}$$

$$\Gamma^p_{\mathbf{Z(i)\,M}}(m) = 1/N \quad \forall\ m \neq 0,\ \Phi(i,k),\ -\phi(i,k)$$

So $\Gamma^p_{\mathbf{Z(i)\,M}}$ has one positive peak at zero delay and $2^i$-1 negative peaks of the same height. If the period of **M** is $N = 2^n-1$, the size of the $i^{th}$ family is

$$K_i = \prod_{j=0}^{i-1} \frac{2^n - 2^{j+1}}{2^i - 2^j} \tag{18}$$

**c)** For our purpose, the chosen family corresponds to i = 2. N has been taken equal to 32 in order to be compared to the above chosen pseudo-orthogonal code. The total number of Y-H sequences is $K_2$ = 140. We have selected M = 32 sequences among them, according to the positions of $\Phi(1)$, $\Phi(2)$, and $\Phi(3)$ on the cross-correlation function (CCF). The family parameters are finally :

$$M = 32\ , \quad N = 31\ , \quad \alpha_R = 5/31$$
$$\Gamma^p{}_{Z(i)\ M}\ (0) = 17/31$$
$$\Gamma^p{}_{Z(i)\ M}\ (\Phi(k)) = -\ 15/31 \qquad k = 1,2,3 \qquad (19)$$
$$\Gamma^p{}_{Z(i)\ M}\ (m) = 1/31 \qquad \forall \quad m \neq 0\ ,\ \Phi(k),\ -\phi(i,k)$$

It is shown **[9]** that $2^{i-1}$ phases are required to identify a Y-H sequence instead of i phases for its construction. Hence, the decision algorithm uses the fact that $2^{i-1}$-1 phases are redudant. Here, i = 2 and two phases are sufficient to identify the sequence. We select the three addresses on the cross-correlation function (CCF) whose amplitudes are the largest in negative. They are ordered according to their value in energy. The first is the largest.

We check then if there exists a sequence characterized by these three phases. If not so, we check one after the other the 2-tuples (1,2),(1,3),(2,3).If we fail to find a sequence, we do not decide. We have considered this situation as an error.

More generally, the same algorithm can be conducted with more than three addresses selected on the CCF. Better results are obtained but the algorithm becomes more and more sophisticated.

## 4) THE UNDERWATER COMMUNICATION EXPERIMENT

In order to test the selected codes in a natural environment, we have conducted an underwater communication experiment in a lake. The receiver was 13.5 m far from the emitter. Both were 4 m deep. This could seem to be an unrealistic experiment, but it has been shown that the channel characteristics, essentially the stability of paths delays, are close to those of an oceanic environment, a medium intensively studied in our laboratory.

The signals were BPSK modulated with a 5 kHz carrier. The occupied bandwidth was 1.25 kHz (half of the main lobe). The noise was generated by an electronic white gaussian noise generator. Two SNR have been tested: 8 dB and -2 dB. They have been measured at the receiver input. The signals have been recorded and then processed in laboratory.

**4.1) Channel identification** : The channel identification shows that there are 3 paths whose strengths and delays are stable. The mean relative strenghs of the other paths to the main one are : 0.73, 0.44 and the mean delays are : 1.53 ms and 3.34 ms. This means that interference exists up to 4 binary digits of each codeword.

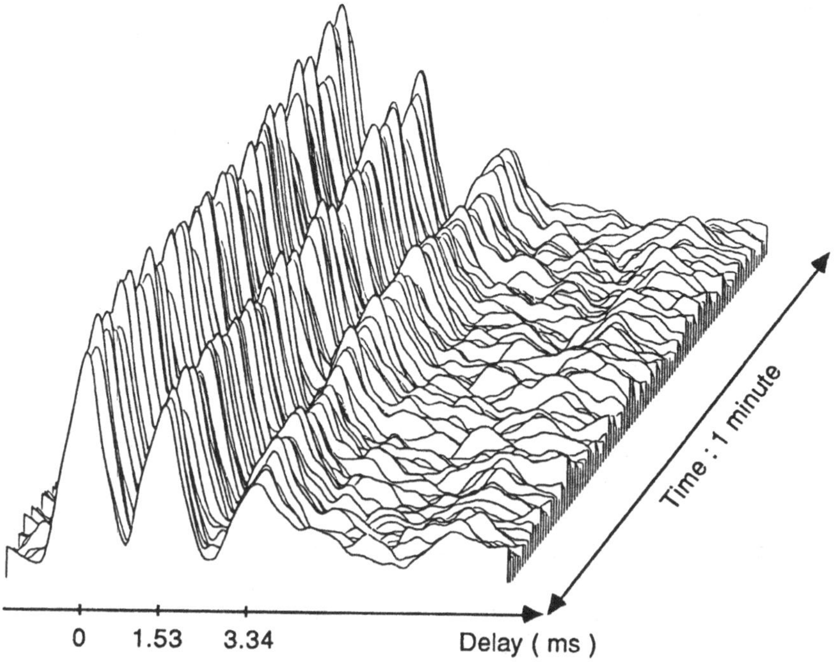


**Figure 3 : Envelope of the channel impulse response**

### 4.2) Continuous channel estimation for optimal processing

The phases and the amplitudes of the different paths are fluctuating all along the transmission. In order to implement receivers using coherent reception scheme, one must perform a continuous channel estimation. The channel estimation is performed under the following hypothesis : the phases and the amplitudes of the different paths keep stable over a codeword duration T. The path delays are constant and known.

We perform a complex demodulation. Correlation is then done on the real and the imaginary part of the signal obtained. Sampling the two CCF at delay $t_j$, one obtains two samples carrying the information on the amplitude and the phase of the path j. This estimation is carried out after each codeword reception.

According to the receiver to be implemented, the amplitudes and/or phasis of one or all paths must be estimated every T.

## 5) RECEIVERS

**5.1) Single-path receiver (receiver 1)** : this receiver works with the signal transmitted through the strongest path. It would be optimal if one deals with a single-path channel.

**5.2) Path energy combining receiver (receiver 2)** : this receiver is optimal with respect to the maximization of the SNR. It combines optimally the signals from all paths. It uses a filter matched to the channel after the first processing given by Fig. 1 and 2.

first processing given by Fig. 1 and 2.

**5.2.1) Processing for Y-H systems** : the positive peak of the CCF $\Gamma^{p}{}_{ZM}$ enables the estimation of the true channel response. Let $\alpha_i$ and $\beta_i$ be respectively the estimated values of the real and the imaginary part of the CCF at delay $t_i$. The impulse response of the channel matched filter is

$$h(t) = \sum_{i=0}^{L-1} (\alpha_i - j\beta_i)\ \delta(t - (\Delta - t_i)) \qquad (20)$$

where $\Delta$ is the channel range spread. It is necessary to avoid that a positive peak issued from a secondary path interfers with a negative characteristic peak of the CCF issued from another path. So we have selected only sequences which possess negative peaks whose adresses on the CCF are more than $\Delta$. In our pratical case, $\Delta$ = 3.34 ms, and we have selected Y-H sequences whose phasis delays are more than $\Phi_{min}$ = 4 ms.

**5.2.2) Processing for the pseudo-orthogonal code** : In the situation of M correlators, one cannot know a priori which one corresponds to the present codeword. Thus, it is impossible to perform directly the channel estimation. However, one can assume that the channel keeps stable over a duration corresponding to two codewords and thus use the previous estimation for the present codeword processing. This hypothesis was not really verified for the phases and we have performed a non-coherent matched filter for receiver 2.

## 6) PERIODICITY SIMULATION FOR Y-H SEQUENCES

The above properties of Y-H correlation impose the sequences are periodicaly repeated. On the other hand, in the underwater experiment, the different Y-H sequences have been transmitted one after another and not repeated. So, it is necessary to simulate the 'periodicity' by cross-correlating the received signal with the parent m-sequence which is itself repeated : so we cross-correlate a signal block of duration T with the parent m-sequence of duration 2T. This processing solves the problem only for the first path.

Let the codewords **X,Y,Z**, successively emitted (**Fig. 4**) :

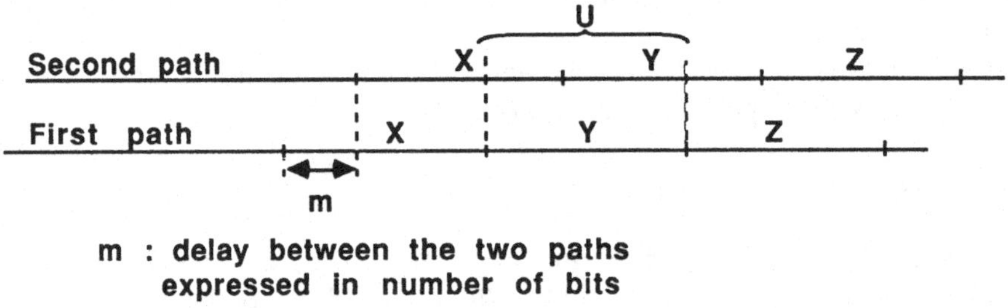


**Figure 4 : Signals configuration**

**U** is a N-block issuing from the second path.

$$\mathbf{U} = ( X_{N-m}, \ldots, X_{N-1}, Y_0, Y_1, \ldots, Y_{N-m+1} )$$

**U** can be written as : $U_r = Y_{r-m} + V_r$ where $\mathbf{V} = (X_{N-m}-Y_{N-m}, \ldots, X_{N-1}-Y_{N-1}, 0, 0, \ldots, 0)$

The correlation is achieved between the following signals

| $a_1 e^{j\theta 1}\mathbf{Y} + a_2 e^{j\theta 2}\mathbf{U}$ | 0 ... 0 |
|---|---|
| **M** | **M** |

where **M** is the parent

This assumes a possible synchronization at the receiver. The CCF leads to, without the additive noise :

$$\Gamma^p_{\mathbf{RM}}(\tau) = a_1 e^{j\theta 1}\,\Gamma^p_{\mathbf{YM}}(\tau) + a_2 e^{j\theta 2}\,\Gamma^p_{\mathbf{YM}}(\tau+m) + a_2 e^{j\theta 2}\,\Gamma^p_{\mathbf{VM}}(\tau) \qquad (21)$$

where the term $\Gamma^p_{\mathbf{VM}}$ is random, depends on the order of emitted Y-H sequences, and will be added to noise. A solution to cancel the term $\Gamma^p_{\mathbf{VM}}$ is the repetition of each Y-H at the emission. But this decreases the transmission rate.

## 7) RESULTS

We have sent $10^4$ codewords for each code. The transmission rate was 1250 bits/s but the real information bit rate was 1250 x $\alpha_R \cong 188$ bit/s. The two receivers have been numerically implemented. The beginning of codewords have been identified by correlation. **Table 1** gives the observed error rates. At 8 dB, the error rate is less than $10^{-4}$ for the multi-correlation codes because no error was observed.

The improvement of the receiver 2 versus receiver 1 is little here because the second and third path amplitudes were all along the transmission smaller than the amplitude of the first one.

The obtained results are in good agreement with the theoretical results in **[9]**.

| SNR | - 2dB | | 8 dB |
|---|---|---|---|
| RECEIVERS | 1 | 2 | 1 |
| PSEUDO | $10^{-1}$ | $4\ 10^{-2}$ | $< 10^{-4}$ |
| YATES | $8\ 10^{-2}$ | $4\ 10^{-2}$ | $6\ 10^{-4}$ |

**Table 1 : Error rates**

## 8) CONCLUSION

This experiment has well carried out the possibility of correct underwater transmissions with small SNR, while keeping reasonable information and error rates.

The pseudo-orthogonal code system appears more robust and more adapted to an unknown channel. On the other hand, it requires more receiving equipment which increases more and more with the number of possible sequences. The Yates-Holgate code system is more sentitive to the repartition of channel path delays, while it remains robust enough. It requires less equipment and is easier to implement ; on the other hand, it allows an easy estimation of the channel. Let us notice that further theoretical comparisons and results between these two kinds of communication systems are presented in **[9]**.

Anyway, the choice of a performant communication system will depend on the available means and the communication characteristics.

**ACKNOWLEDGMENTS**. *This work has been partly supported by the Direction of the French Naval Constructions.*

## REFERENCES

[1] F. CHAVAND, *Transmission d'Information dans les Canaux Multitrajets à caractéristiques Aléatoires par Codage Pseudo-Orthogonal. Application au Canal Ionosphérique.* Thèse d'Etat, Orsay, 1981.

[2] G. HAKIZIMANA, G. JOURDAIN, and G. LOUBET, *Coding for Communication through Multipath Channels and Application to Underwater Case.* Proc. EUSIPCO, pp. 1087-1090 The Hague, The Netherlands, Sept. 1986.

[3] G. HAKIZIMANA, *Etude de familles de séquences binaires.* Rapport CEPHAG n° 44/85.

[4] G. HOFFMAN DE VISME, *Binary Sequences.* English Univ. Press, London,1971, pp. 43-48.

[5] J.M. POTTER, *Recursive Code Generation Based on m-Sequence.* Electronics Letters Vol.16, n° 22, pp. 858-859, 1980.

[6] D. SARWATE and M.B. PURSLEY, *Crosscorrelation Properties of Pseudo-Random and Related Sequences.* Proc. IEEE, Vol. 68, n° 5, pp. 593-619, 1980.

[7] G.L. TURIN, *Introduction to Spread-Spectrum Antimutipath Techniques and their Application to Urban Digital Radio.* Proc. IEEE, Vol. 68, n° 3, pp. 328-353, 1980.

[8] K.W. YATES and D.J. HOLGATE, *Code Modulation of m-Sequence.* Electronics Letters Vol. 15, pp. 836-838, 1979.

[9] G. HAKIZIMANA, *Codages adaptés aux communications avec trajets multiples,* Thèse de Doctorat, INPG, Grenoble, Sept. 1987.

# CODING AND DECODING ALGORITHMS OF REED-SOLOMON CODES EXECUTED ON A M68000 MICROPROCESSOR

Francisco J. García-Ugalde
División de Estudios de Posgrado
Facultad de Ingeniería, UNAM
A.P. 70-256 C. Universitaria
04510 MEXICO, D.F.

## ABSTRACT

This paper presents a comparison of different algebraic methods for decoding Reed-Solomon (RS) codes. Three kinds of decoding algorithms have been studied, two of them are <<time domain>> algorithms, while the third employs a transform decoding method. We achieve with this development, measurements that will permit the selection of the method with optimal decoding in the case of one future microprogramming implementation. Optimal decoding implies that the faster algorithm is the best one. It can be concluded that acceptable RS decoders can be constructed with microprocessors only for low speed transmission rates.

## RESUME

Cet article présente une comparaison des différentes méthodes algébriques pour le décodage des codes de Reed Solomon (RS), Nous avons étudié trois sortes d'algorithmes de décodage. Parmi eux, deux sont définis dans le <<domaine du temps>>, et le troisième utilise une méthode de décodage par transformée. Avec ce développement nous disposons d'un moyen qui nous permettra de sélectioner, dans le cas d'une future réalisation microprogrammée, la méthode de décodage optimal dans le sens de sa rapidité. Nous concluons qu'on peut construire des décodeurs de Reed-Solomon avec des microprocesseurs seulement pour des debits de transmission suffisament faibles.

Indexing terms: Cyclic codes, error-correcting codes, Reed-Solomon codes, simulation of Reed-Solomon codes, decoding algorithms.

## 1. INTRODUCTION

In digital communications systems Reed-Solomon (RS) codes have been largely used. Many applications have been developed for spatial communications and data transmission systems [1] - [5] . Principally, this is due to their powerful error correction capacity and their algebraic structure that has permited the development of high performance algorithms for coding as well as decoding [6]-[8] . These codes can correct both random and burst errors. Nevertheless the delay introduced by computing several steps for coding and principally for decoding is one of their major disadvantages [9] .

Therefore in this work the complexity associated to coding and decoding of RS codes for different algorithms, and different values of parameters has been analysed. The algorithms associated to coding and decoding have been executed in assembly language on a M68000 microprocessor, utilizing as efficiently as possible the internal microprocessor architecture. As a previous step for a future realization these simulations allow direct measurements of execution times, associated with the algorithms, in order to compare them.

Thus the complexity has been measured proportionally, to the execution time of the algorithms, utilizing a 16-bit microprocessor in its version of 4 MHz and N-channel technology. Three kinds of decoding algorithms have been studied, two of them are <<time domain>> algorithms [10],[11], while the third one employs a transform decoding method [7]. The execution time for encoding to compare it with decoding times has also been given.

Even though the M68000 was chosen to test the algorithms, it should be emphasized that the primary interest is not in the absolute performance on this microprocessor. The results obtained with the M68000 and the comparisons made can be extrapolated to the future generation of microprocessors such that the potential performance of these algorithms on faster, more capable microprocessors can be revealed.

## 2. GENERALITIES

The structure of RS codes, and standard methods for encoding and decoding is briefly described. The standard method for decoding is then compared to the other variants of algebraic decoding utilized in this work.

Figure 1 shows an schematic RS encoder where information symbols are grouped into blocks k symbols long. To each of these k symbol blocks, n-k redundant symbols are added to produce an n symbol code word with $n > k$. The code used is defined over the Galois field $GF(2^m)$. Each symbol of the code is a field element in $GF(2^m)$ where $GF(2^m)$ denotes the finite field of $2^m$ symbols. Hence the number of bits per symbol is equal to m.

The $n-k = d-1$ redundant symbols are used for error correction where d is the minimum distance of the code. For RS codes d is also equal to $d = 2t+1$, and t denotes the number of errors that may be corrected in each codeword. The 2t redundant symbols for a systematic RS code can be obtained as the coefficients of the remainder of

$$x^{n-k}\ i(x)/g(x) \tag{1}$$

where $i(x)$ is the information polynomial whose coefficients are the k information symbols, thus

$$i(x) = \sum_{\ell=0}^{k-1} i_\ell\, x^\ell = i_o + i_1\, x + i_2 x^2 + \ldots + i_{k-1} x^{k-1} \quad \text{where } i_\ell \in GF(2^m) \tag{2}$$

$g(x)$ is the generator polynomial of the code [12] , defined by

$$g(x) = \prod_{i=j}^{j+2t-1} (x-\alpha^i) = \sum_{\ell=0}^{2t} g_\ell\, x^\ell \tag{3}$$

where j is a specified nonnegative integer, often chosen to be 1; $\alpha$ is a primitive element in $GF(2^m)$, and $g_\ell \in GF(2^m)$ are the coefficients of $g(x)$ with $g_{2t}=1$. The generator polynomial defined above does not have symmetrical coefficients, i.e.,

$$g_\ell \neq g_{2t-\ell} \qquad \text{for} \quad 0 \leq \ell \leq 2t$$

except when $j = 2^{m-1} - t$. In the latter case,

$$g_\ell = g_{2t-\ell} \qquad \text{for} \quad 0 \leq \ell \leq 2t \tag{4}$$

and

$g_o = g_{2t} = 1$

Note that in the latter case, only t multipliers are needed in an encoder. Thus using this generator polynomial will reduce the number of multipliers required to implement RS encoders by one half [13],[14]. Nevertheless, for simplicity let us consider the case where $j = 1$.

The codeword obtained $c(x)$, corresponds to a systematic encoding, i. e., it is formed by a first part which is equal to the information symbols

$$c_\ell = i_{\ell-n+k} \qquad n-1 \geq \ell \geq n-k \tag{5}$$

and a second part which is equal to the redundant symbols, i.e.

$$c_\ell = r_\ell \qquad n-k-1 \geq \ell \geq 0 \tag{6}$$

where $c_\ell$ are the symbols of the codeword also considered to be the coefficients of the polynomial $c(x)$, and $r_\ell$ are the coefficients of the polynomial $r(x)$. Thus, the transmitted codeword is given by

$$c(x) = x^{n-k}\, i(x) + r(x) = \sum_{\ell=0}^{n-1} c_\ell\, x^\ell = c_o + c_1 x + \ldots + c_{n-1} x^{n-1} \tag{7}$$

From (1) and (7), note that we have $c(x) = g(x) \cdot q(x)$, i.e., the codeword must be also a multiple of $g(x)$.

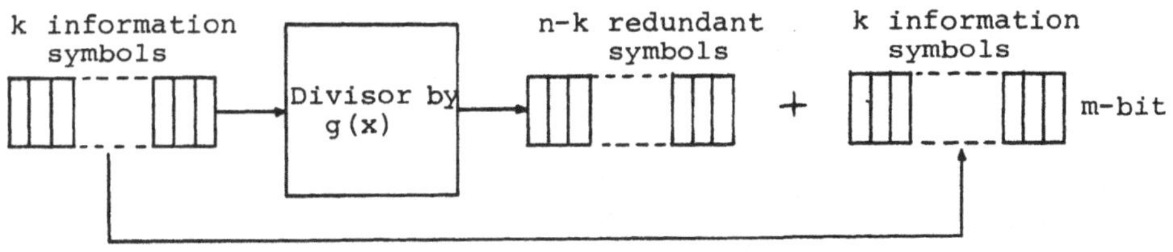


Figure 1 Schematic RS encoder.

## Decoding

After transmission over a noisy channel, the received word is

$$R(x) = \sum_{\ell=0}^{n-1} R_\ell x^\ell = R_o + R_1 x + \ldots + R_{n-1} x^{n-1} \quad \text{where } R_\ell \in GF(2^m) \qquad (8)$$

Then the error pattern added by the channel is

$$E(x) = R(x) - c(x) = \sum_{\ell=0}^{n-1} E_\ell x^\ell = E_o + E_1 x + \ldots + E_{n-1} x^{n-1} \qquad (9)$$

where

$$E_\ell = R_\ell - c_\ell \quad \text{and} \quad E_\ell \in GF(2^m) \qquad 0 \leq \ell \leq n-1 \qquad (10)$$

Because an RS code of block length n and minimum distance d can be used to correct up to t errors where $2t+1 \leq d$. The function of the decoder is to find the codeword which has the minimum distance from the received word. If the number of errors is greater than t, either the decoder fails to decode and the error pattern is detected, or the decoder decodes incorrectly and an undetected post-decoder error pattern results. Detected error patterns can often be dealt with either by flagging or suppressing the unreliable data. Figure 2 shows that incorrect decoding occurs if the error pattern in the received word is decoded to a codeword other than the transmitted one. This occurs when the error pattern is located within a sphere of radious t (in an n dimension space) surrounding a codeword other than the transmitted one. In such a case, the error pattern is decoded to the codeword at the center of the sphere [15].

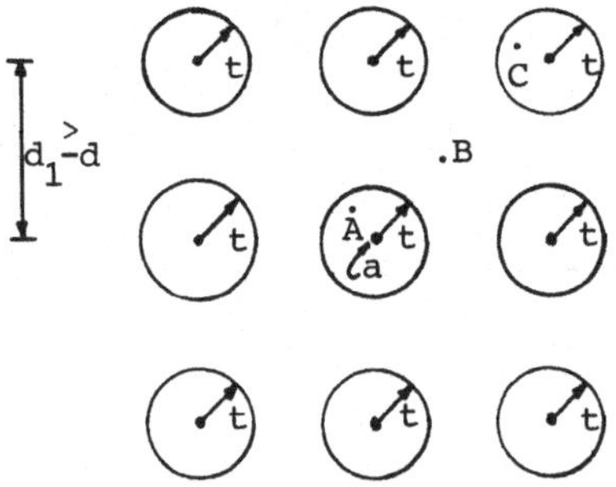


Figure 2 Schematic representation of decoding.
a. Transmitted codeword, A. Received word which yields a correct decoding, B. Received word which yields a detected impossible decoding, C. Received word which yields an undetected incorrect decoding.

In some situations it is desirable for the decoder to correct erasures in addition to errors. The above analysis can be extended to include the case of errors and erasures decoding, considering the effect of deleting symbol positions in a RS code. This is essentially what happens when a symbol is erased. To the decoder, an erasure is a symbol in a received word that has been labelled "unreliable" by the demodulator, i.e., it is an error whose location is known before correction begins and whose magnitud may or may not be zero [6]. Reed-Solomon codes can be used to fill erasures as well as correct errors. The algorithm presented is to correct successfully patterns of e errors and f erasures of the word of length n, as long as $2e+f<d$. Henceforth, the relationship between the transmitted codeword c(x) and the received word R(x) is

$$R(x) = c(x) + E(x) + F(x) \tag{11}$$

where

$$E(x) = \sum_{\ell=0}^{n-1} E_\ell x^\ell \quad \text{and} \quad F(x) = \sum_{\ell=0}^{n-1} F_\ell x^\ell \tag{12}$$

E(x) is the error polynomial and F(x) is the erasure polynomial, where $E_\ell$ and $F_\ell \in GF(2^m)$.

The first step of the decoding procedure is to store the received word R(x) into a buffer register and then compute the syndrome components $S_i$ using the equation

$$S_i = R(\alpha^i) = \sum_{j=0}^{n-1} R_j \, \alpha^{ij} \qquad 1 \le i \le 2t \tag{13}$$

By (11), $S_i$ can be written as

$$S_i = c(\alpha^i) + E(\alpha^i) + F(\alpha^i) \tag{14}$$

Since c(x) is a multiple of g(x), whose roots are $\alpha^i$ for $1 \le i \le 2t$, the above equation can be expressed as

$$S_i = E(\alpha^i) + F(\alpha^i) = \sum_{j=0}^{n-1} E_j\alpha^{ij} + \sum_{k=0}^{n-1} F_k\alpha^{ik} \qquad 1 \le i \le 2t \tag{15}$$

Let $Y_j$ be the j-th error magnitude, and $X_j=\alpha^j$ be the j-th error location. And let $Z_k$ be the k-th erasure magnitude and $W_k=\alpha^k$ be the k-th erasure location, where $X_j \ne W_k$. Then,

$$E_i = \sum_{j=0}^{n-1} E_j \alpha^{ij} = \sum_{j=1}^{e} Y_j X_j^i \qquad 1 \le i \le 2t \tag{16}$$

and

$$F_i = \sum_{k=0}^{n-1} F_k \alpha^{ik} = \sum_{k=1}^{f} Z_k W_k^i \qquad 1 \le i \le 2t \tag{17}$$

Hence, by (16) and (17), $S_i$ in (15) can be written as

$$S_i = E_i + F_i = \sum_{j=1}^{e} Y_j X_j^i + \sum_{k=1}^{f} Z_k W_k^i \qquad 1 \le i \le 2t \tag{18}$$

The erasure locator polynomial $\tau(x)$ is defined as

$$\tau(x) = \prod_{k=1}^{f} (1 - W_k x) = \sum_{\ell=0}^{f} \tau_\ell x^\ell = 1 + \sum_{\ell=1}^{f} \tau_\ell x^\ell \tag{19}$$

The coefficients $\tau_\ell$'s are elementary symmetric functions of $W_k$. Since

$$\tau[W_k^{-1}] = 0 = 1 + \sum_{\ell=1}^{f} \tau_\ell W_k^{-\ell} \qquad 1 \le k \le f \tag{20}$$

the roots of $\tau(x)$ are the inverse erasure locations.

Forney polynomial $T(x)$ is calculated from the polynomial $\tau(x)$ and the syndrome polynomial

$$S(x) = \sum_{i=1}^{2t} S_i x^i \tag{21}$$

in the following manner

$$T(x) = (1 + S(x))\tau(x) - 1 \quad \text{modulo} \quad x^{2t+1} = \sum_{i=1}^{2t} T_i x^i \tag{22}$$

The $T_i$'s are called Forney syndromes or modified cyclic parity cheks. Then the second and third steps of the decoding procedure are respectively, to compute $\tau_\ell$ for $1 \le \ell \le f$ and Forney syndromes.

The error location polynomial [16] $\sigma(x)$ is defined as

$$\sigma(x) = \prod_{j=1}^{e} (1 - X_j x) = \sum_{\ell=0}^{e} \sigma_\ell x^\ell = 1 + \sum_{\ell=1}^{e} \sigma_\ell x^\ell \tag{23}$$

Similarly, $\sigma_\ell$'s are elementary symmetric functions of $X_j$. Since

$$\sigma(X_j^{-1}) = 0 = 1 + \sum_{\ell=1}^{e} \sigma_\ell \, X_j^{-\ell} \qquad 1 \leq j \leq e \tag{24}$$

the roots of $\sigma(x)$ are the inverse error locations.

By (18), (20), (22) and (24) it is demonstrated [16] that the coefficients $\sigma_\ell$'s, satisfy

$$T_{i+f+e} + \sum_{\ell=1}^{e} T_{i+f+e-\ell} \, \sigma_\ell = 0 \qquad 1 \leq i \leq 2t - f - e \tag{25}$$

Hence the fourth step of the decoding procedure is to compute $\sigma_\ell$ for $1 \leq \ell \leq e$ in (25) from the Forney syndromes. This can be accomplished by using Berlekamp's iterative algorithm [16] or Massey's linear feedback shift register (LFSR) synthesis algorithm [17].

Upon obtaining the $\sigma_\ell$'s, the fifth step of the decoding procedure is to compute the roots of the error location polynomial by a Chien search [6]. It consists merely of substituting all n possible nonzero error locations, i.e., for $x = X_j^{-1} = \alpha^{-j}$ for $0 \leq j \leq n-1$. If $\sigma(\alpha^{-j}) = 0$, then $j$ represents the error location.

Knowing the error locations, step sixth is to compute the erasure and error magnitudes, which is done by calculating first the erasure-and-error evaluator polynomial.

$$\Omega(x) = (1 + T(x))\sigma(x) \quad \text{modulo} \quad x^{2t+1} \tag{26}$$

Thus the erasure-and-error magnitudes can be computed from the formula

$$B_j = \frac{A_j^{(e+f-1)} \; \Omega(A_j^{-1})}{\prod_{j \neq i} (A_j - A_i)} \qquad 1 \leq j \leq e + f \tag{27}$$

Where the $A_j$'s are the erasure-and-error locations. If $A_j$ is equal to an error location, i.e., $A_j = X_j$, then $B_j$ is the corresponding error magnitude, i.e., $B_j = Y_j$. On the other hand, if $A_j$ is equal to an erasure location, i.e., $A_j = W_k$, then $B_j$ is the corresponding erasure magnitude, i.e., $B_j = Z_k$.

Then the corrected codeword is obtained by subtracting both the error polynomial $E(x)$ and the erasure polynomial $F(x)$ from the stored received word $R(x)$ in the buffer register.

Note that if f=0, by (19), $\tau(x)=1$. Thus by (22), $T_i=S_i$, and the $\sigma_\ell^{\prime}$s can be computed from the Berlekamp's or Massey's algorithm with the syndrome components $S_i^{\prime}$s, instead of the Forney syndromes $T_i^{\prime}$s. By [16], in this case the coefficients $\sigma_\ell^{\prime}$s satisfy

$$S_{i+e}+\sum_{\ell=1}^{e} S_{i+e-\ell}\,\sigma_\ell=0 \qquad 1 \leq i \leq 2t-e \tag{28}$$

On the other hand if $e=0$, by (23), $\sigma(x)=1$. Thus (26) is reduced to

$$\Omega(x) = (1+T(x)) \qquad \text{modulo} \quad x^{2t}+1 \tag{29}$$

## 2. VARIANTS OF THE ALGEBRAIC DECODING OF REED-SOLOMON CODES

The other two algorithms for decoding RS codes that have been utilized, are described in the form of flowcharts. Afterwards, these two algorithms will be compared with the algorithm described in the preceding section.

Figure 3 shows another algorithm for decoding RS codes which also operates in the <<time domain>>. This one is different from the one described previously because the Berlekamp's iterative algorithm is inseminated directly with the erasure locator polynomial $\tau(x)$ [10]. Thus the polynomial P(x)

$$P(x) = \tau(x)\,\sigma(x) \tag{30}$$

called product polynomial or combined erasure-and-error locator polynomial is issued from this step.

The other substantial difference from this algorithm with regard to the algorithm of the preceding section is that for computing the erasure-and-error magnitudes, the divisor of (27) is replaced by the derivative of P(x) evaluated at $A_j$ [11], where the $A_j^{\prime}$s are the erasure-and-error locations.

Figure 4 shows the third algorithm utilized, which employs a transform decoding method, described in [7]. The transform method is used to compute the syndrome components $S_i$ for $1 \leq i \leq 2t$, and the inverse transform of the syndrome vector $(S_o,S_1,\ldots S_{n-2})$ to obtain the concatenated erasure-and-error polynomial defined by

$$V(x) = E(x) + F(x)$$

where E(x) and F(x) have been defined by (12).

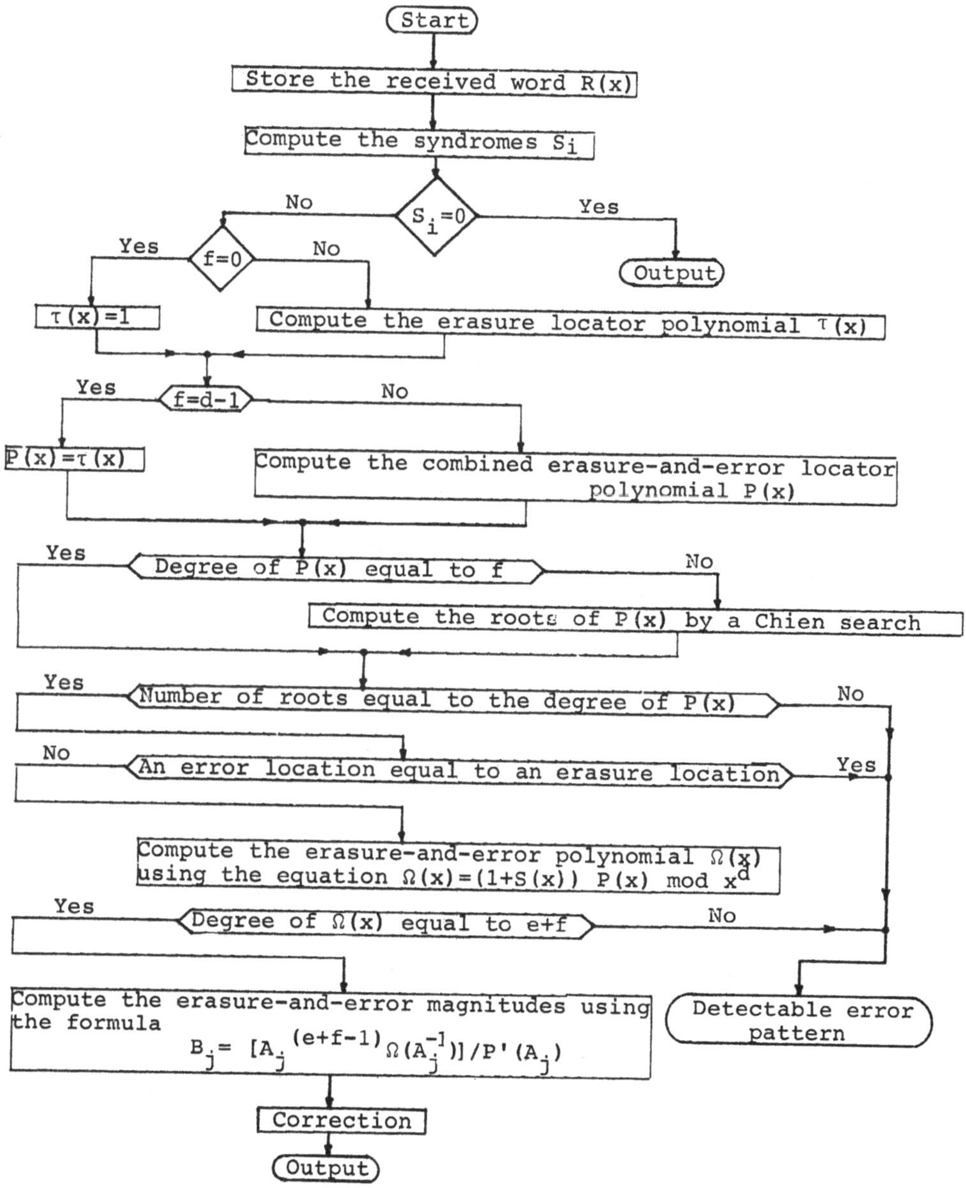


Figure 3 Flowchart of the algorithm 2 for decoding RS codes operating in the <<time domain>>.

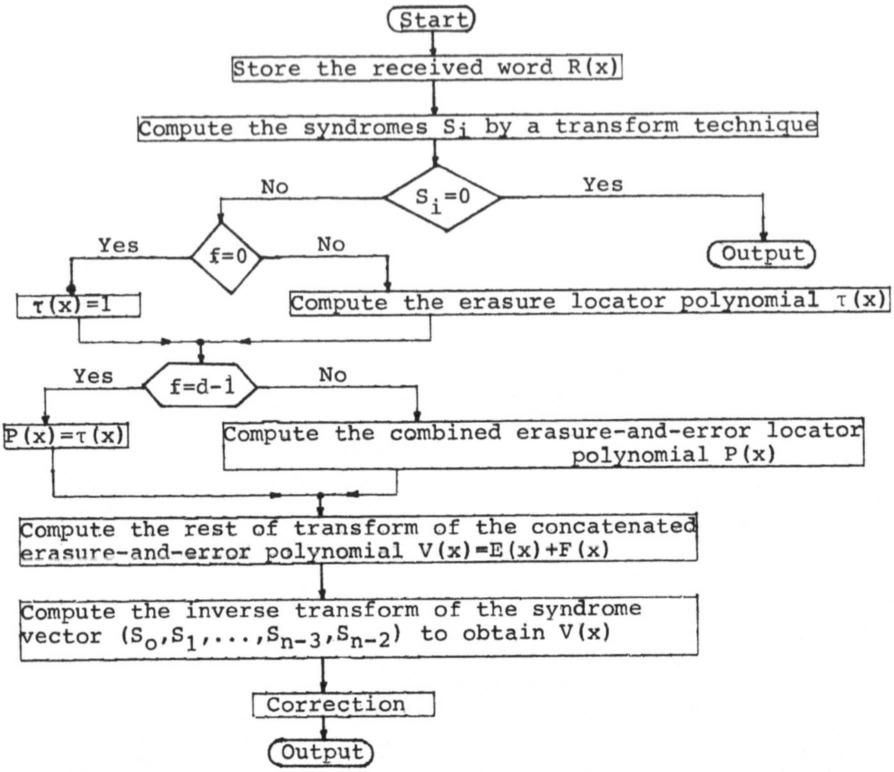


Figure 4  Flowchart of the algorithm 3 for decoding RS codes utilizing a transform decoding method.

A sufficient number of trials to obtain reliable average values of the execution times of the three described algorithms has been accomplished. A random variable with uniform distribution has been utilized to generate both erasure and error locations and magnitudes.

## 3. MEASURED RESULTS

Figures 5 and 6 show the average execution times matrices for decoding RS codes with parameters (15,9,7) and (31,25,7) respectively. Only average values have been preserved because changes on execution times have been detected in the function of erasure and error locations. Maximal devation from the average values, and average execution time for encoding have also been given.

| e \ f | 0 | 1 | 2 | 3 | 4 | 5 | 6 | Curve |
|---|---|---|---|---|---|---|---|---|
| 0 | 20 | 23.7 | 27 | 30.5 | 34 | 36.8 | 40.1 | T115 e=0 |
| 1 | 29.2 | 33.1 | 36.8 | 40.9 | 44 | | | T115 e=1 |
| 2 | 35 | 41.1 | 46.4 | | | | | T115 e=2 |
| 3 | 46.6 | | | | | | | T115 e=3 |

(a) Standard method

| e \ f | 0 | 1 | 2 | 3 | 4 | 5 | 6 | Curve |
|---|---|---|---|---|---|---|---|---|
| 0 | 23 | 26.4 | 29.7 | 33.2 | 36.3 | 39.9 | 42.5 | T215 e=0 |
| 1 | 29.4 | 35.7 | 40.8 | 47.5 | 55 | | | T215 e=1 |
| 2 | 37 | 45 | 54.1 | | | | | T215 e=2 |
| 3 | 43 | | | | | | | T215 e=3 |

(b) Algorithm 2 for decoding in the <<time domain>>

| e \ f | 0 | 1 | 2 | 3 | 4 | 5 | 6 | Curve |
|---|---|---|---|---|---|---|---|---|
| 0 | 29 | 30.3 | 31.5 | 32.6 | 33.8 | 35 | 36.2 | T315 e=0 |
| 1 | 32.5 | 34.3 | 35.8 | 37.5 | 39 | | | T315 e=1 |
| 2 | 37 | 39.2 | 41.5 | | | | | T315 e=2 |
| 3 | 42 | | | | | | | T315 e=3 |

(c) Algorithm 3 utilizing a transform decoding method

Figure 5 Decoding execution time matrices for the RS(15,9,7) code Maximal deviation=±3.5 Average encoding time=6.7 Time in miliseconds, e=number of errors, f=number of erasures.

| e \ f | 0 | 1 | 2 | 3 | 4 | 5 | 6 | Curve |
|---|---|---|---|---|---|---|---|---|
| 0 | 45 | 49.3 | 53 | 57.4 | 60.7 | 65 | 69.1 | T131 e=0 |
| 1 | 56.2 | 60 | 63.8 | 68.5 | 72.5 | | | T131 e=1 |
| 2 | 66.2 | 71 | 75.7 | | | | | T131 e=2 |
| 3 | 83.2 | | | | | | | T131 e=3 |

(a) Standard method

| e \ f | 0 | 1 | 2 | 3 | 4 | 5 | 6 | Curve |
|---|---|---|---|---|---|---|---|---|
| 0 | 47.5 | 51.6 | 55.4 | 59 | 62.8 | 67.1 | 71 | T231 e=0 |
| 1 | 56 | 62.4 | 70 | 76.7 | 84.5 | | | T231 e=1 |
| 2 | 66 | 75.2 | 86 | | | | | T231 e=2 |
| 3 | 79.5 | | | | | | | T231 e=3 |

(b) Algorithm 2 for decoding in the <<time domain>>

| e \ f | 0 | 1 | 2 | 3 | 4 | 5 | 6 | Curve |
|---|---|---|---|---|---|---|---|---|
| 0 | 55 | 56.3 | 57.5 | 59 | 60 | 61.3 | 62.6 | T331 e=0 |
| 1 | 59.4 | 61.1 | 63.4 | 65.6 | 67.6 | | | T331 e=1 |
| 2 | 69.1 | 71 | 72.7 | | | | | T331 e=2 |
| 3 | 77.6 | | | | | | | T331 e=3 |

(c) Algorithm 3 utilizing a transform decoding method

Figure 6 Decoding execution time matrices for the RS(31,25,7) code Maximal deviation = ±4.5 Average encoding time = 27.5 Time in miliseconds, e=number of errors, f=number of erasures.

The results obtained are plotted in figures 7 and 8 to compare the three algorithms utilized.

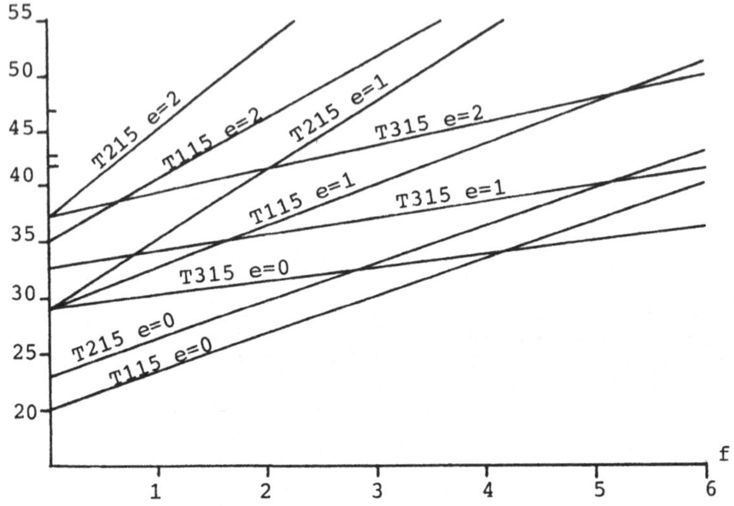


Figure 7 Decoding execution times for the RS(15,9,7) code

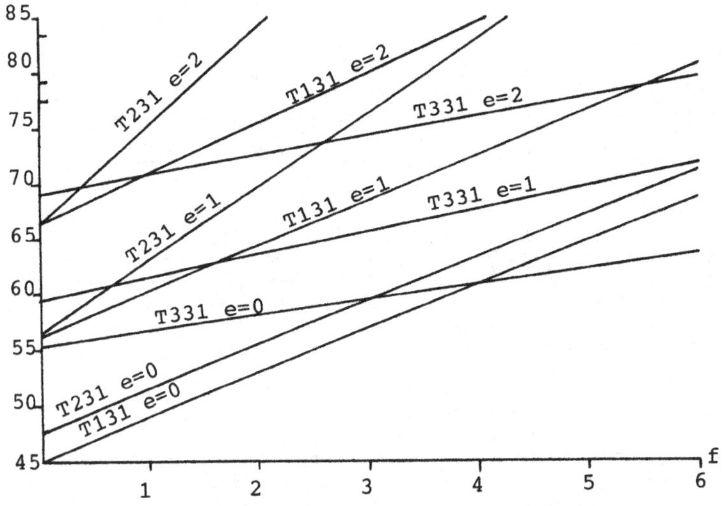


Figure 8 Decoding execution times for the RS(31,25,7) code

## 4. CONCLUSION

This work has permited the comparison of three decoding algorithms for Reed-Solomon codes by executing them in a 16-bit microprocessor.

Figures 7 and 8 show that decoding algorithms that employ a transform decoding method have a comportment more independent from the number of erasures and errors. This is the result of computing the syndromes $S_i$ by a transform technique and computing the inverse transform of the syndrome vector $(S_o, S_1, \ldots, S_{n-3}, S_{n-2})$ to obtain the concatenated erasure-and-error polynomial V(x) which are both independent of t(the number of errors that may be corrected in each codeword).

The second algorithm which operates also in the <<time domain>> has a divergent comportment with respect to the standard decoding procedure. This is due to the Chien search because in this case, when the number of erasures f is not equal to zero then the number of roots of the product polynomial P(x) defined by (30), must be superior to the number of roots of the error locator polynomial $\sigma(x)$ taking alone.

In the execution of the three algorithms tested, internal microprocessor architecture has been exploited at a maximum, i.e., utilizing as best as possible both address and data internal registers, as soon as the extensive number of addressing modes. Nevertheless the execution times are still important. Thus the use of microprocessors to implement a Reed-Solomon decoder is justified only for low data transmission rates.

## REFERENCES


[1] R.F. Rice, "Channel coding and data compression system considerations for efficient communication of planetary imaging data", Tech. Memo. 33-695, Jet Propulsion Laboratory, Pasadena, CA, June 1974.

[2] A. Hauptschein, "Practical, high performance concatenated coded spread spectrum channel for JTIDS", in Proc. Nat. Telecommun. Conf. 1977, p. 35: 4-1 to 4-8.

[3] K.Y. Liu and J. Lee, "An experimental study of the concatenated Reed-Solomon/Viterbi channel coding system performance and its impact on space communications", in Proc.Nat. Telecommun. Conf., 1981.

[4] K.Y. Liu and K.T. Woo, "The effects on receiver tracking phase error on the performance of concatenated Reed-Solomon/Viterbi channel coding system", in Proc. Nat.Telecommun. Conf., 1980, pp. 51.5.1-51.5.5.

[5] Odenwalder et al, "Hybrid coding system study", submited to NASA Ames Research Center Linkabit Co. San Diego, CA, Final Rep., Contract NAS-2-6722, Sept. 1972.

[6] W.W. Peterson and E.J. Weldon, Jr, Error-Correcting Codes, Cambridge MA: MIT Press, 1972.

[7] R.L. Miller, T.K. Truong and I. S. Reed, "Efficient program for decoding the (255,223) Reed-Solomon code over $GF(2^8)$ with both errors and erasures, using transform decoding", IEE Proc. Vol. 127, Pt. E, No.4, July 1980.

[8] I.S. Reed, T.K. Truong and R.L. Miller, "Simplified algorithm for

correcting both errors and erasures of Reed-Solomon codes", IEE Proc., Vol. 126, No.10, October 1979.

[9] E. Lenormand, "Réalisation d'un codeur-décodeur (31,15) de Reed-Solomon", Revue Technique Thomson-CSF, Vol.12,No.3,September 1980.

[10] R.E. Blahut, "Transform techniques for error control codes", IBM J. Res. Develop., Vol.23, No.3, May 1979.

[11] I.S. Reed, T.K. Truong and R.L. Miller, "Decoding of B.C.H. and R. S. codes with errors and erasures using continued fractions", Electronics Letters, Vol.15, No.17, 16th August 1979.

[12] F. Mac Williams and N. Sloane, The Theory of Error-Correcting Codes, Amsterdam: North-Holland Publishing Co., 1977.

[13] In-Shek Hsu, I.S. Reed, T.K. Truong, Ke Wang, Chiunn-Shyong Yeh and L.J. Deutsch, "The VLSI implementation of a Reed-Solomon encoder using Berlekamp's bit-serial multiplier algorithm", IEEE Trans. on Computers,Vol.C-33, No.10, October 1984.

[14] Kuang Yung Liu, "Architecture for VLSI design of Reed-Solomon decoders", IEEE Trans. on Computers, Vol.C-33, No.2, February 1984.

[15] F.J. García Ugalde, "Les performances des codes de Reed-Solomon sur un canal discret sans memoire", X Colloque sur le Traitement du Signal et ses Applications, Nice du 20 au 24 Mai 1985, pp.619-624.

[16] E.R. Berlekamp, Algebraic Coding Theory, New York, McGraw-Hill, 1968.

[17] J.L. Massey, "Shift-register synthesis and BCH decoding", IEEE Trans. on Inform. Theory, Vol.IT-15, pp.122-127, Jan. 1969.

Vol. 270: E. Börger (Ed.), Computation Theory and Logic. IX, 442 pages. 1987.

Vol. 271: D. Snyers, A. Thayse, From Logic Design to Logic Programming. IV, 125 pages. 1987.

Vol. 272: P. Treleaven, M. Vanneschi (Eds.), Future Parallel Computers. Proceedings, 1986. V, 492 pages. 1987.

Vol. 273: J.S. Royer, A Connotational Theory of Program Structure. V, 186 pages. 1987.

Vol. 274: G. Kahn (Ed.), Functional Programming Languages and Computer Architecture. Proceedings. VI, 470 pages. 1987.

Vol. 275: A.N. Habermann, U. Montanari (Eds.), System Development and Ada. Proceedings, 1986. V, 305 pages. 1987.

Vol. 276: J. Bézivin, J.-M. Hullot, P. Cointe, H. Lieberman (Eds.), ECOOP '87. European Conference on Object-Oriented Programming. Proceedings. VI, 273 pages. 1987.

Vol. 277: B. Benninghofen, S. Kemmerich, M.M. Richter, Systems of Reductions. X, 265 pages. 1987.

Vol. 278: L. Budach, R.G. Bukharajev, O.B. Lupanov (Eds.), Fundamentals of Computation Theory. Proceedings, 1987. XIV, 505 pages. 1987.

Vol. 279: J.H. Fasel, R.M. Keller (Eds.), Graph Reduction. Proceedings, 1986. XVI, 450 pages. 1987.

Vol. 280: M. Venturini Zilli (Ed.), Mathematical Models for the Semantics of Parallelism. Proceedings, 1986. V, 231 pages. 1987.

Vol. 281: A. Kelemenová, J. Kelemen (Eds.), Trends, Techniques, and Problems in Theoretical Computer Science. Proceedings, 1986. VI, 213 pages. 1987.

Vol. 282: P. Gorny, M.J. Tauber (Eds.), Visualization in Programming. Proceedings, 1986. VII, 210 pages. 1987.

Vol. 283: D.H. Pitt, A. Poigné, D.E. Rydeheard (Eds.), Category Theory and Computer Science. Proceedings, 1987. V, 300 pages. 1987.

Vol. 284: A. Kündig, R.E. Bührer, J. Dähler (Eds.), Embedded Systems. Proceedings, 1986. V, 207 pages. 1987.

Vol. 285: C. Delgado Kloos, Semantics of Digital Circuits. IX, 124 pages. 1987.

Vol. 286: B. Bouchon, R.R. Yager (Eds.), Uncertainty in Knowledge-Based Systems. Proceedings, 1986. VII, 405 pages. 1987.

Vol. 287: K.V. Nori (Ed.), Foundations of Software Technology and Theoretical Computer Science. Proceedings, 1987. IX, 540 pages. 1987.

Vol. 288: A. Blikle, MetaSoft Primer. XIII, 140 pages. 1987.

Vol. 289: H.K. Nichols, D. Simpson (Eds.), ESEC '87. 1st European Software Engineering Conference. Proceedings, 1987. XII, 404 pages. 1987.

Vol. 290: T.X. Bui, Co-oP A Group Decision Support System for Cooperative Multiple Criteria Group Decision Making. XIII, 250 pages. 1987.

Vol. 291: H. Ehrig, M. Nagl, G. Rozenberg, A. Rosenfeld (Eds.), Graph-Grammars and Their Application to Computer Science. VIII, 609 pages. 1987.

Vol. 292: The Munich Project CIP. Volume II: The Program Transformation System CIP-S. By the CIP System Group. VIII, 522 pages. 1987.

Vol. 293: C. Pomerance (Ed.), Advances in Cryptology – CRYPTO '87. Proceedings. X, 463 pages. 1988.

Vol. 294: R. Cori, M. Wirsing (Eds.), STACS 88. Proceedings, 1988. IX, 404 pages. 1988.

Vol. 295: R. Dierstein, D. Müller-Wichards, H.-M. Wacker (Eds.), Parallel Computing in Science and Engineering. Proceedings, 1987. V, 185 pages. 1988.

Vol. 296: R. Janßen (Ed.), Trends in Computer Algebra. Proceedings, 1987. V, 197 pages. 1988.

Vol. 297: E.N. Houstis, T.S. Papatheodorou, C.D. Polychronopoulos (Eds.), Supercomputing. Proceedings, 1987. X, 1093 pages. 1988.

Vol. 298: M. Main, A. Melton, M. Mislove, D. Schmidt (Eds.), Mathematical Foundations of Programming Language Semantics. Proceedings, 1987. VIII, 637 pages. 1988.

Vol. 299: M. Dauchet, M. Nivat (Eds.), CAAP '88. Proceedings, 1988. VI, 304 pages. 1988.

Vol. 300: H. Ganzinger (Ed.), ESOP '88. Proceedings, 1988. VI, 381 pages. 1988.

Vol. 301: J. Kittler (Ed.), Pattern Recognition. Proceedings, 1988. VII, 668 pages. 1988.

Vol. 302: D.M. Yellin, Attribute Grammar Inversion and Source-to-source Translation. VIII, 176 pages. 1988.

Vol. 303: J.W. Schmidt, S. Ceri, M. Missikoff (Eds.), Advances in Database Technology – EDBT '88. X, 620 pages. 1988.

Vol. 304: W.L. Price, D. Chaum (Eds.), Advances in Cryptology – EUROCRYPT '87. Proceedings, 1987. VII, 314 pages. 1988.

Vol. 305: J. Biskup, J. Demetrovics, J. Paredaens, B. Thalheim (Eds.), MFDBS 87. Proceedings, 1987. V, 247 pages. 1988.

Vol. 306: M. Boscarol, L. Carlucci Aiello, G. Levi (Eds.), Foundations of Logic and Functional Programming. Proceedings, 1986. V, 218 pages. 1988.

Vol. 307: Th. Beth, M. Clausen (Eds.), Applicable Algebra, Error-Correcting Codes, Combinatorics and Computer Algebra. Proceedings, 1986. VI, 215 pages. 1988.

Vol. 309: J. Nehmer (Ed.), Experiences with Distributed Systems. Proceedings, 1987. VI, 292 pages. 1988.

Vol. 310: E. Lusk, R. Overbeek (Eds.), 9th International Conference on Automated Deduction. Proceedings, 1988. X, 775 pages. 1988.

Vol. 311: G. Cohen, P. Godlewski (Eds.), Coding Theory and Applications. Proceedings, 1986. XIV, 196 pages. 1988.